Essential Windows Phone 7.5

Essential Windows Phone 7.5

Application Development with Silverlight

■ Shawn Wildermuth

✦✦ Addison-Wesley

Upper Saddle River, NJ • Boston • Indianapolis • San Francisco
New York • Toronto • Montreal • London • Munich • Paris • Madrid
Capetown • Sydney • Tokyo • Singapore • Mexico City

Many of the designations used by manufacturers and sellers to distinguish their products are claimed as trademarks. Where those designations appear in this book, and the publisher was aware of a trademark claim, the designations have been printed with initial capital letters or in all capitals.

The .NET logo is either a registered trademark or trademark of Microsoft Corporation in the United States and/or other countries and is used under license from Microsoft.

Microsoft, Windows, Visual Basic, Visual C#, and Visual C++ are either registered trademarks or trademarks of Microsoft Corporation in the U.S.A. and/or other countries/regions.

The author and publisher have taken care in the preparation of this book, but make no expressed or implied warranty of any kind and assume no responsibility for errors or omissions. No liability is assumed for incidental or consequential damages in connection with or arising out of the use of the information or programs contained herein.

The publisher offers excellent discounts on this book when ordered in quantity for bulk purchases or special sales, which may include electronic versions and/or custom covers and content particular to your business, training goals, marketing focus, and branding interests. For more information, please contact:

U.S. Corporate and Government Sales
(800) 382-3419
corpsales@pearsontechgroup.com

For sales outside the United States, please contact:

International Sales
international@pearson.com

Visit us on the Web: informit.com/aw

Library of Congress Cataloging-in-Publication Data

Wildermuth, Shawn.
  Essential windows phone 7.5 : application development with silverlight
/ Shawn Wildermuth.
        p.   cm.
  Includes index.
  ISBN 978-0-321-75213-0 (pbk. : alk. paper)
1. Windows phone (Computer file) 2. Silverlight (Electronic resource)
3. Operating systems (Computers) 4. Application software—Development.
5. Mobile computing—Programming. I. Title.
  QA76.59.W54 2012
  005.4'46—dc23
                                              2011036842

ISBN-13: 978-0-321-75213-0
ISBN-10:   0-321-75213-9
Text printed in the United States on recycled paper at RR Donnelley in Crawfordsville, Indiana.
First printing, December 2011

To my friend and mentor, Chris Sells,
without whom I would have never learned
that the story is more important than the facts.

Contents at a Glance

Contents

Figures

Tables

Foreword

When Shawn asked me to write a foreword for his Windows Phone development book, I had a couple of reactions. First, that they must really be scraping the bottom of the barrel if they have asked me to write anything. There are so many people who actually help bring the product to market who never really get the credit they deserve. While I am honored that I was asked to write this, based in part on my public role on the team, the engineering team that designed and built this amazing product are the real heroes. The product itself is amazing, but the right application platform, which enables the amazing Metro apps and games to be built, is a developer's playground. I do this to honor them.

My second reaction was to think about the huge value Shawn has in the Microsoft ecosystem. As an eight-time MVP and Silverlight Insider, Shawn's contributions are highly valued both for their content as well as for their reach. When Shawn speaks, you know that he has the developer in mind: He is a developer's developer. Without individuals like Shawn, it would be tough (if possible at all) for Microsoft to have built our developer ecosystem over the last three decades. I do this to honor him.

My last reaction was one of panic. I have never written a foreword before, so I was at a bit of a loss as to what I should say. I figure if you are buying this book, you did so of your own volition, and not on the strength of what I have to say here. However, if you are reading the foreword with an eye toward confirming your belief that Windows Phone is where it's at, well, for that I can be accommodating. I do this to honor you.

With the initial release of Windows Phone, and the subsequent pairing with Nokia, Microsoft is investing in building the third ecosystem for mobile developers. The canvas with which mobile developers can work on Windows Phone is unlike any other platform, whereby developers can create simply gorgeous apps with more focus on the user experience than tinkering with the innards of a convoluted framework. Metro apps come alive on the screen, and you will be able to build deeply engaging applications using Live Tiles.

Windows Phone 7.5 is an updated release, codenamed "Mango," and carries with it the tagline "Put people first." We think the same way about the developer platform. We aim to put developers first. The book you are holding might be your first step on your journey to building Windows Phone apps. It may be a refresher course. Either way, with Shawn's guidance, we know that you will come away from this experience feeling great about your prospects of building amazing mobile experiences for Windows Phone, and a firm belief that Microsoft puts the developers first when we think about Windows Phone. Every developer matters. Every. Single. One.

—Brandon Watson
   Microsoft Corporation

Preface

I have never owned a PalmPilot. But I have owned palmtops and smart-phones. I dived into writing software for a plethora of different devices but never got very far. My problem was that the story of getting software onto the phones was chaotic and I didn't see how the marketing of software for phones would lead to a successful product. In the intervening years, I got distracted by Silverlight and Web development. I didn't pay attention as the smartphone revolution happened. I was happily neck-deep in data binding, business application development, and teaching XAML.

The smartphone revolution clearly started with the iPhone. What I find interesting is that the iPhone is really about the App Store, not the phone. It's a great device, but the App Store is what changed everything, providing a simple way to publish, market, and monetize applications for these handheld powerhouses that everyone wanted. Of course, Apple didn't mean to do it. When the original iPhone shipped, Apple clearly said that Safari (its Web browser) was the development environment. With the pressure of its OS X developer community, Apple relented and somewhat accidentally created the app revolution.

When it was clear that I had missed something, I dived headlong into looking at development for phones again. I had an Android phone at the time, so that is where I started. Getting up to speed with Eclipse and Java wasn't too hard, but developing for the phone was still a bit of a chore. The development tools just didn't seem to be as easy as the development I was used to with Visual Studio and Blend. In this same time frame, I grabbed a Mac and tried my hand at Objective-C and Xcode to write something simple for the iPhone. That experience left me bloodied and bandaged. I wanted to write apps, but since it was a side effort, the friction of the tool sets for Android and iPhone left me wanting, and I put them aside.

Soon after my experience with iPhone and Android, Microsoft took the covers off its new phone platform: Windows Phone 7. For me, the real excitement was the development experience. At that point I'd been teaching and writing about Silverlight since it was called WPF/E, so the ability to marry my interest in mobile development to my Silverlight knowledge seemed like a perfect match.

I've enjoyed taking the desktop/Web Silverlight experience I have and applying the same concepts to the phone. By being able to use Visual Studio and Blend to craft beautiful user interface designs and quickly go from prototype to finished application, I have found that the workflow of using these tools and XAML makes the path of building my own applications much easier than on other platforms.

In the middle of this learning process Microsoft continued to mature the platform by announcing and releasing Windows Phone 7.5 (code-named Mango). I was left questioning whether to finish my Windows Phone 7 book or rush forward and mold all the new features of Windows Phone 7.5 into a book for this next version of the phone. Obviously you know the answer to that question.

It has been a long road to get the right story for this book, and to help both beginners and existing Silverlight developers to learn from the book. My goal was always to allow readers to get started writing apps quickly, while also including the information that leads to great apps. Because of the relative size of these minicomputers we keep in our pockets, knowing when to pull back is often the key to a great application. As you will see throughout this book, my goal has been to help you build great apps, not rich applications. This means I will try to hold your hand as you read the book, but I will also challenge your assumptions about how you approach the process of building applications for the phone.

Acknowledgments

Writing a book is a team sport. Anyone who thinks for a moment that writing a book requires that you sit in a dark room and craft words that magically get bound into Amazon currency hasn't been through the sausage factory that is book writing. The fact is that I may have the skills to get words down on virtual paper, but I am not good at much of the rest of the process. It takes a strong editor who knows how to dole out praise and pressure in equal amounts. It takes technical reviewers who aren't afraid to ruffle your feathers. It takes production people to take the mess of Visio ramblings you call figures and create something the reader will understand. Finally, it takes an army of people to listen to your questions about the ambiguity of writing a book based on a beta version of a product . . . and who will not stop responding to your constant pestering. So I'd like to thank my army of people by acknowledging their real contributions (in no particular order).

First and foremost, I want to thank my editor at Addison-Wesley, Joan Murray. I am not an easy author to work with, and she's been a trouper in getting me to stick to deadlines and coercing me to make the right decisions, not just the easy ones. The rest of the people at Addison-Wesley that I've had the pleasure to work with are all great, too. Of special note, Christopher Cleveland did a great job picking up the role of developmental editor in the middle of the book, and has been great through the whole process.

To the litany of people on the Silverlight Insiders Mailing List and the Windows Phone 7 Advisors Mailing List, I would like to thank you for your patience as I pestered the lists with endless questions and hyperbolic rants. You all helped shape this book, even if you didn't realize it.

During this process, my blog's readers and my followers on Facebook and Twitter remained a consistent sounding board. My polls and open questions helped me shape what is and isn't in this book. For that I am indebted to you.

I also want to thank my terrific technical reviewers, Jeremy Likeness, Ambrose Little, and Bruce Little. Not only did they help me find the tons of places I just plain got it wrong, but they also helped me when the story got off track and I missed that key piece of the puzzle. Of particular note, I want to thank Ambrose for his tenacious adherence to the designer's voice. He helped me make sure I wasn't coddling the developers into bad user experience design.

To anyone else I forgot to mention, I apologize.

—Shawn Wildermuth
   November 2011
   http://wildermuth.com
   @shawnwildermuth

About the Author

During his twenty-five years in software development, **Shawn Wilder-muth** has experienced a litany of shifts in software development. These shifts have shaped how he understands technology. Shawn is a nine-time Microsoft MVP, a member of the INETA Speaker's Bureau, and an author of several books on .NET. He is also involved with Microsoft as a Silverlight Insider and a Data Insider. He has spoken at a variety of international conferences, including TechEd, MIX, VSLive, OreDev, SDC, WinDev, DevTeach, DevConnections, and DevReach. He has written dozens of articles for a variety of magazines and websites including *MSDN*, DevSource, InformIT, *CoDe Magazine*, ServerSide.NET, and MSDN Online. He is currently teaching workshops around the United States through his training company, AgiliTrain (http://agilitrain.com).

1

Introducing Windows Phone

To some, the cell phone is an annoying necessity; to others, it's a critical need. Being able to use a phone to make calls everywhere has really changed the way people communicate. In the past few years these phones have taken another leap forward. With the introduction of iPhone and Android devices, the consumer market for an always-connected device that can interact with the Internet, run applications, and make phone calls has changed people's relationship with their phone. It has also raised the bar for consumer-level devices. Consumers now expect their phones to also function as GPSs, gaming devices, and Internet tablets. For some consumers, their phones are now their primary connections to the Internet, replacing the desktop/laptop computer for the first time. As developers, our challenge is to find the best way to create the experiences the user needs. Windows Phone provides the platform, and Silverlight is the engine to power those experiences.

A Different Kind of Phone

When Microsoft originally unveiled Windows Phone 7 many skeptics expected the phone would simply try to play catch-up with Apple's and Google's offerings. Microsoft had other plans, though. The new operating system for the phone was a departure from existing offerings from the other mobile operating system vendors (primarily Apple, Research in Motion, and Google). Instead of just mimicking the icon pattern screens that iPhone and Android seemed to love, Microsoft thought in a different way. Application and operating system design is defined in a new design language code-named Metro.[1] This design language defines a set of guidelines and styles for creating Windows Phone applications. The design of the Start screen laid out by Metro is similar to other smartphone designs in that it is a list of icons. Instead of separating the icons into pages, Windows Phone lets users scroll through the icons. Windows Phone is also differentiated from other smartphones in that each icon can include information about the application. These icons are called **Live Tiles,** as shown in Figure 1.1.

What Is a Design Language?

Developers think about a language as a set of textual expressions that describe some machine operation(s). For designers, it is a set of rules for defining the look and feel of a set of applications (or an entire operating system in this case). Wikipedia.org defines it more generally as "*. . . an overarching scheme or style that guides the design of a complement of products or architectural settings.*"

The Start screen should be a place where users can quickly review the status of the phone. The Live Tiles will give user information such as the number of missed phone calls and the number of email or SMS messages waiting, or even third-party information such as the current weather. When you develop your own applications you can either create a simple icon for the Start screen or build a Live Tile for your users.

For applications, the Windows Phone screen is divided into three areas in which the user can interact with the phone: the system tray, the logical client, and the application bar (see Figure 1.2).

The system tray area is managed by the phone's operating system. This is where the time, signal strength, and alerts will appear to the user. Most applications will leave this area of the screen visible to the user. Some

---

1. UI Design and Interaction Guide for Windows Phone: http://shawnw.me/wpmetroguide

FIGURE 1.1 Windows Phone Start screen

FIGURE 1.2 Phone screen real estate

applications (e.g., games) may hide this area, but you should only do so when critical to the success of your application.

The logical client area is where your application will exist. This area shows your user interface and any data and points of interaction.

The application bar shows options for your application. While using the application bar is not a requirement, it is a very common practice as it gives users access to your application's options and menus. For example, Figure 1.3 shows a simple note-taking application that uses the application bar to allow users to create new notes or show the menu (note that the ellipsis can be clicked to open the list of menu items).

One big distinction that users will see in many of the applications built into Windows Phone is the use of **hubs.** The central idea of a hub is to provide a starting point to get the user to use natural curiosity to learn what is available in the application. Usually these hubs take the form of applications that are larger than the phone screen. Instead of the typical page-based

FIGURE 1.3 The application bar in action

applications that are fairly commonplace on smartphones, the Metro style guide introduces something called a **panorama application.** For panorama applications the phone is used as a window that looks into a larger application surface. You'll notice in Figure 1.4 that the content of the screen takes up most of the horizontal real estate, but the next section of the panorama application shows up on the right side of the screen to help the user understand that there is more content.

As the user navigates through the panorama application, the virtual space is moved within the window. For example, in Figure 1.5 you can see how, after sliding the application to the left, the rightmost part of the panorama becomes visible.

Figure 1.4 Panorama application

Figure 1.5 Last pane of a panorama application

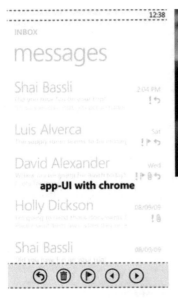
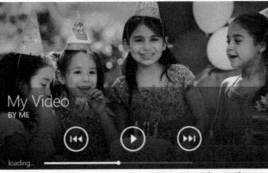

app-UI with chrome   **app-UI without chrome**

FIGURE 1.6 Using Metro chrome, or not

The use of the panorama application results in a simple but powerful user interface design that users should find very intuitive.

By following the guidelines specified by Metro, you can create applications that should be consistent with the rest of the phone, while giving you the freedom to create applications of any kind. In this way, Metro helps by defining basic ideas of how a Windows Phone application should look so that the user can see complete consistency. At the same time, Metro says you can simply take over the entire user interface and not use the basic ideas of the Metro chrome, leaving you the flexibility to create either custom experiences or applications that look like they belong on the phone.

Figure 1.6 shows example apps with and without the chrome applied.

Integrated Experiences

One of the main purposes of the phone is to be an integrated platform on which applications can interact with each other and with the core phone experiences. This means you can write applications that integrate with the phone in unprecedented ways. Table 1.1 outlines some of the core experiences users will be able to interact with and use on the phone.

TABLE 1.1 Integrated Experiences

Experience	Description
People	The people on your phone, including contacts and past phone calls/SMS messages
Office	Integration with email, Word, Excel, and PowerPoint files
Music+Videos	The media on your device
Marketplace	Access to try, buy, and install applications on the phone
Pictures	View, share, and take pictures on your device
Games	Playing games on the device; this includes Xbox Live integration

As developers, your code might or might not look like the code in traditional applications. Because you can write straightforward applications that can be launched in a traditional sense, the integrated experiences let your applications also interact with and even be embedded into these experiences. This means you can write applications that extend and power these experiences.

Phone Specifications

For Windows Phone, the stakes were high in terms of Microsoft's ability to not only create the software but also encourage its partners to build the phone. Learning a lesson from its past Windows Mobile platform, Microsoft decided to be very specific about the hardware to ensure a great user experience while giving phone designers some flexibility with feature sets so that they could compete with one another. Table 1.2 shows the hardware requirements.

In addition, Windows Phone has physical requirements. The most obvious of these is that each phone must have seven standard inputs, as shown in Figure 1.7.

Table 1.3 lists and describes these seven hardware inputs.

TABLE 1.2 Hardware Specifications

Category	Requirement
Screen resolution	WVGA (480 x 800)
Capacitive touch	At least four points of touch support
Memory	256MB RAM, 8GB Flash
Sensors	A-GPS, Accelerometer, Compass, Light and Proximity, Gyro
CPU	ARM7 Scorpion/Cortex or better (typically 1GHz+)
GPU	DirectX 9 acceleration
Camera	5 megapixels minimum, flash required
Bluetooth	Bluetooth 2.1 + EDR; Bluetooth profiles provided are Hand-Free Profile (HFP), Headset Profile (HSP), Advanced Audio Distribution Profile (A2DP), and Phone Book Access Profile (PBAP)
Multimedia	Codec acceleration required; support for DivX 4, 5, and 6 as well as H.264 High Profile required (High Profile is used by Blu-ray)
Wi-Fi	802.11g radio required
Radio	FM radio receiver required

FIGURE 1.7 Seven points of input

TABLE 1.3 Hardware Inputs

	Input	Expected Behavior
1	Power button	When powered off, a long press will power on the device. If powered on and screen is active, will turn off screen and lock device. If screen is off, will enable screen and present unlock UI.
2	Volume control	A rocker switch will adjust volume for current activity's sound profile (e.g., phone call volume while on a call). Pressing volume during a phone call will disable the ringer. Adjusting volume when no activity is presently happening will allow user to switch between sound profiles.
3	Touch screen	The capacitive touch screen will support at least four points of touch.
4	Camera button	A long press on this dedicated button will launch the camera application.
5	Back button	This button issues a "back" operation. This may take the user back in an individual application or from one application to the previous application as presented by the page API.
6	Start button	This takes the user to the Start screen of the device.
7	Search button	This launches the search experience to allow searching across the device.

Now that we have seen what the phone consists of, let's see how users will interact with the phone.

Input Patterns

You are the developer. You want users to want to use your applications. That means you must deal with the different ways the phone can accept user input. Developing for the Web or desktop means you are primarily dealing with designing for the keyboard and mouse. But when developing for the phone, you have to change the way you look at input and consider that the user is going to interact with your application in different ways. Interaction patterns for the phone include touch, keyboards (hardware and software), hardware buttons, and sensors.

Designing for Touch

The Metro design language is specifically constructed to make sure the interface is treating touch as a first-class citizen and that the interface requires no training (i.e., is intuitive). By building a design language that defines the elements of a touch-based interface, Microsoft has made it easier to build such interfaces. The design language includes guidelines for what touch gestures are supported, as well as how to space and size elements for finger-size interactions. Figure 1.8 shows an example from Metro to define the minimum sizes for touch points and their spacing.

Metro also defines the types of interactions (e.g., touch gestures) the device supports. Most of these interactions are well-worn gestures that have been the vocabulary of other touch devices such as the iPhone, Zune HD, and Android. These interactions include

- Single touch:
  - Tap
  - Double-Tap
  - Pan
  - Flick
  - Touch and Hold
- Multitouch:
  - Pinch/Stretch/Rotate

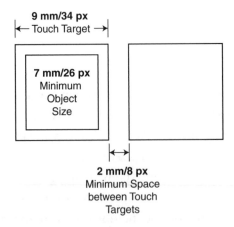

FIGURE 1.8 Metro's interactive element sizes

In addition to specifying the types of gestures, Metro specifies the use cases for each interaction. For example, the Double-Tap interaction is specifically used to zoom in and zoom out. This use case is explicitly different from what the typical desktop Windows developer might expect. But for the sake of consistency, Metro maps out what the user should expect with these interactions. While Metro would not be read by actual users, it would be the basis for the phone's built-in application. Interactions of your applications should match the rest of the phone, therefore adhering to the principle of least surprise for the user. This also hints at the reality that the phone design is not supposed to be based on users' expectations of how Windows works, but be more obvious than that. The touch interaction is much different from a mouse, and the overall hope (as far as I can tell) is to help users get a feel for the right interaction without training them.

Hardware Buttons

Windows Phone requires that each phone has three hardware buttons on the front of the device. As described in the "Phone Specifications" section earlier in this chapter, these three buttons have discrete actions. The only one you really need to concern yourself with is the Back button. Not only should the Back button move the user from your application to the last running application (the default behavior), it should also allow the user to move from state to state in your application. As you develop applications for the phone, be aware of what the user might expect from the Back button. Taking advantage of this can make your application even more intuitive.

Keyboards

Since not all interactions will be simple gestures but must be able to support text entry, the UI Design and Interaction Guide stipulates that a software keyboard (or Soft Input Panel or SIP) should be available for every text entry (as even in a keyboarded phone, users should be able to type on the screen). Keyboards are provided by the operating system by default. As users attempt to edit text (e.g., the user taps on a text box) the operating system displays a software keyboard to enable touch-based keyboard entry. Figure 1.9 shows the default keyboard.

FIGURE **1.9 Default keyboard**

The Metro style guide also specifies that the keyboards should be con-
textually relevant depending on the type of text to be typed. For example,
Figure 1.10 shows an email keyboard and a phone number keyboard.

While there are a number of layouts, Metro specifies a few specifically,
as shown in Table 1.4.

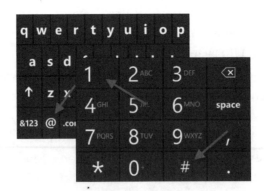

FIGURE **1.10 Contextual keyboards**

TABLE 1.4 Sample Keyboard Layouts

Keyboard	Description
Default	Standard QWERTY keyboard
Text	Includes autocorrect panel
Chat	Includes an emoticon key and autocorrect
Email address	Includes .com and @ keys
Phone number	12-key numeric layout
Web address	Includes .com key and "Go" key, which instructs the application that user input has completed
Maps	Includes "Go" key, which instructs the application that user input has completed
SMS address	Simplified layout with quick access to phone numbers

The purpose of Metro is to help developers (both externally and at Microsoft) create applications that make the platform feel cohesive and consistent.

Sensors

You should consider that not all the input to the phone is typical. It is important that you, the application developer, open your mind to different types of input. Windows Phone supports a number of sensors that will allow you to take input in these different forms (see Table 1.5).

TABLE 1.5 Sensors

Sensor	Description
Accelerometer	Detects the position of the phone in three dimensions, as well as movement such as shaking or tilting
Compass	Determines the direction that the phone is facing in relation to the magnetic poles of the Earth
Proximity	Determines how close someone is to the face of the phone

continues

TABLE 1.5 Sensors (*continued*)

Sensor	Description
Light	Determines the amount of ambient light around the phone
Gyro	Detects the active rotation of the phone in three dimensions
A-GPS	Determines the location of the phone on the physical face of the Earth (e.g., longitude and latitude)

Application Lifecycle

The user experience is the most important feature in Windows Phone. After learning many lessons from the competition and from its own experience with Windows Mobile devices, Microsoft decided it would control process execution on the phone. The main reason for this is that on a device like this, the number of applications running can severely impact the quality of the user experience. On the Windows Mobile platform and Android devices full multitasking is allowed, but most users quickly learn to use a task-killer application to kill applications that no longer are required to be opened. This is an adequate solution for multitasking but does require that the memory on the device be managed by users. While power users will be comfortable with this, most users will not.

To enable developers to build rich applications that act and feel as though multitasking is enabled, Windows Phone uses an approach that allows applications to be paused, made dormant, and suspended without having to alert the user that the application is being paused. It does this by notifying the application when it is being paused; then the application is also notified when it is to resume running. In the pause and resume states, the application is given a chance to save and load data to give the user the impression that the application never stopped. In Figure 1.11 you can see how an application will go through the five states during its lifetime. This lifecycle is called **tombstoning.**

If you ignore the pause and resume states, your application will simply act as though it was restarted by the operating system. This lifecycle is

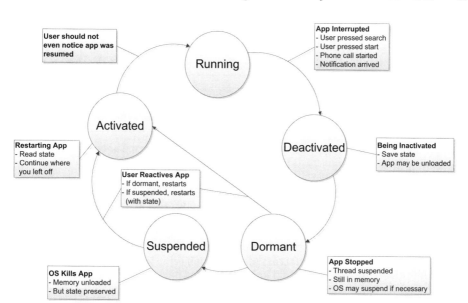

FIGURE 1.11 Application lifecycle (tombstoning)

used for the majority of applications. Microsoft allows only a small number of partners to run outside this lifecycle.

Driving Your Development with Services

While some applications will only access data on the phone, in practice many applications will need to use the data connectivity to interact with servers and the cloud.[2] The phone is a connected device (meaning Internet connectivity is available most of the time). This means you can power your applications via traditional services such as Web APIs or Web services. These are typically services that are either Web-enabled (like Amazon's Web APIs) or custom services you write in the cloud.

To power the phone, Microsoft has also exposed a number of services to simplify phone development, as described in Table 1.6.

---

2. In this book, when I say "cloud" I am referring to Internet-based services including services on your own hardware, services in cloud-based hosting (e.g., Azure, Amazon EC2), or public Web APIs (e.g., Amazon's Web APIs). It does not mean your services have to be hosted in a cloud-based solution.

TABLE 1.6 Microsoft Phone Services

Sensor	Description
Location	Accesses information about the location of the phone. Uses GPS if available; otherwise, uses other location-based information (e.g., cell towers).
Notifications	Supports sending asynchronous data to the phone. Typically end up as toast notifications on the phone that can launch your application or updating of Live Tiles.
Xbox Live	Access to user's Gamerscore and other game information stored on Microsoft servers. Also allows game developers to grant Gamerscore to players.
App Deployment	Provides access to information in the Marketplace, and supports trial (or try before you buy) purchases and update management of applications. Also includes support to tell users what rights the application is requesting.

Live Tiles

The center of the entire user interface paradigm in Windows Phone 7 is the notion of the Start screen. The hub is the main screen that users will be presented when they boot up or turn on the phone. Unlike the interface that the two main competitors (iPhone and Android devices) present, the hub is not just a collection of application icons, but rather a set of Live Tiles. These tiles include information about the state of the information inside the application. For example, the People tile in the Start screen will include pictures of the last few updates the people on your device have had. This is an indication that you may want to go to the People application on your phone to see the updates. Figure 1.12 shows this transition from tile to application.

This lets you, the application developer, control what the tile looks like. So you could decide on just a simple numbering system like the Phone or Outlook tiles, or you could change the look and feel completely, like the People tile.

The way that tiles get updated is powerful as well. Ordinarily you might consider that applications themselves would update the tiles, but that would assume your application would need to be launched whenever

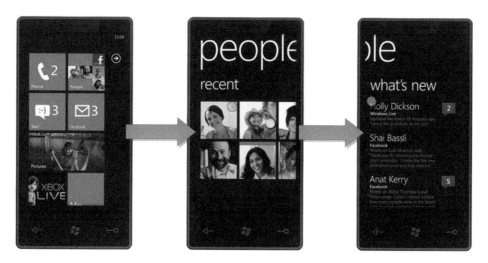

FIGURE 1.12 A tile in the hub

the tile needs updating. Instead, the phone uses the Notification Service to allow you to send an update to the phone through Microsoft's own service to update the tile. This works well because the update is very performant (as the update simply includes the information about the updated tile and never needs to start your application to update the tile). In addition, this means that ordinarily cloud services (like server-side applications) can update the tile as well in a very efficient way. Figure 1.13 illustrates a simple update to a tile.

While it is easy to treat the hub like the Desktop in Windows, it should be treated differently. It's the dashboard to the user's data, not just shortcuts to applications. Users should be able to view the hub and see the basic information they need in order to decide how to interact with the data. For example, if there are new email messages, voice-mail messages, and Facebook updates, users should be able to see at a glance what is happening to themselves online and allow their phone to be a window to that world.

FIGURE 1.13 Updating tiles

The Marketplace

In this post-iPhone world, it's all about the apps. Unlike earlier iterations of Microsoft phone technologies (e.g., Windows Mobile), all software will have to go through Microsoft to get installed on the device. While this may upset longtime Windows Mobile developers, it is something we have always needed on Microsoft phones. Users need a place to find good applications from a source that can guarantee that the software will not interfere with the device's ability to run smoothly. That's where the Marketplace comes in.

Distributing Your Application through the Marketplace

As a developer, you do not have a choice of how to distribute your application; you must use the Marketplace. The Marketplace is a partnership between you and Microsoft. It enables you to deliver your application with very little work on your part. For Microsoft's part, the Marketplace does the following:

- Handles billing via credit card or operator billing (i.e., bill to provider)
- Gives you a 70% revenue share
- Allows your apps to be updated without cost, regardless of whether it is a paid, free, or ad-supported app
- Lets you deliver trial versions of applications and convert them on-the-fly to full versions
- Handles automatic updating of your application

To distribute through the Marketplace you have to join the Marketplace so that it can validate who you are and set up revenue sharing. It costs $99 per year to join the Marketplace, and membership allows you to submit 100 free apps per year (without incurring extra costs) and an unlimited number of paid apps. With Marketplace membership you can also register up to five phones as development phones. Microsoft wants Marketplace membership to enable you to be successful, as the more applications you sell, the more money both you and Microsoft make.

Marketplace Submissions

The Marketplace allows developers to submit applications to be handled through an approval process. The brunt of the approval process consists of certification testing to ensure that the application does not violate the rules of the Marketplace. Figure 1.14 illustrates the approval process.

The process starts with you creating your application and packaging it as a .xap[3] file. At that point you would go to the Marketplace website and submit the application. Microsoft then verifies that the .xap file is valid and asks you to enter additional metadata (e.g., the publisher information,

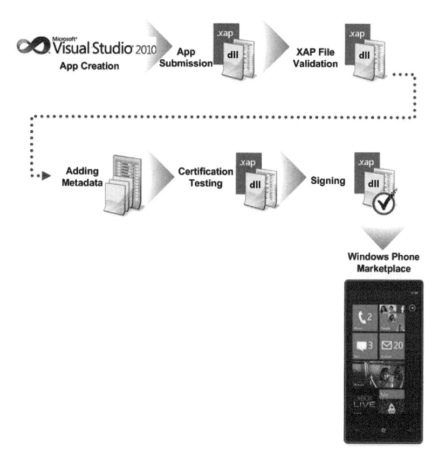

FIGURE 1.14 Marketplace application submission process

---

3. .xap is a packaging format for Silverlight and XNA applications. It is simply a ZIP file with all the code and assets that are needed to run the application.

a short description, etc.). Next, Microsoft runs the application through a certification process to check the quality of the software (to make sure it does not destabilize the phone) and check that the application follows its Marketplace policies (explained shortly). Finally, if the application passes certification, Microsoft signs the .xap file and posts it on the Marketplace to allow it to be sold and/or downloaded.

But what does certification testing really mean? This step of the process is part quality assurance and part content filtering. The quality assurance part of the process is to ensure that your application is stable and does not affect the reliability of the phone. The quality assurance part is based on the following criteria.

- Your application must run on any Windows Phone regardless of model, keyboard hardware, and manufacturer.
- Your application must shut down gracefully even in the event of unhandled exceptions.
- Your application must not hang or become unresponsive.
- Your application must render within five seconds of launch and be responsive within 20 seconds.
- Your application must be able to resume after being paused without losing functionality or data.
- Your application must handle access to the Back button correctly: Back on the first page should exit the application; Back on subsequent pages should correctly move backward in the application.
- Your application must not interfere with phone calls or SMS/MMS messaging in any way.

In addition to checking the quality of the application, the certification also checks the application for adherence to the policies of the Marketplace. There are two types of policies: Application Policies and Content Policies.

Application Policies

Microsoft has detailed a set of policies that every application must adhere to. These policies are meant to protect the users, Microsoft, and the phone carriers. The application requirements can be divided into basic policies that govern the size and functionality of your applications, and legal policies that stop bad people from doing things that will hurt the carriers, Microsoft, or the users. The current policies as of the writing of this book include

- Basic application policies
- Legal usage policies

Basic Application Policies

A number of standard policies apply to application size, functionality, and usage.

- Your application must be fully functional when acquired from the Windows Phone Marketplace except if additional data is required to be downloaded as permitted (see below).
- The size of your application cannot exceed 500MB. If you want to enable installation over-the-air (OTA), the size must not exceed 20MB. Larger applications will be available to be downloaded via Wi-Fi or a tethered connection.
- If your application requires the download of a large additional data package (e.g., more than 50MB) to enable the application to run as described, the application description must disclose the approximate size of the data package and that additional charges may apply depending on connectivity used to acquire data.
- If your application includes a trial version, the trial version must reasonably represent the functionality and quality of the full application.

Legal Usage Policies

Additionally, specific policies are related to payments, legality, and use of personal information.

- Your application may not require the user to pay outside of the Windows Phone Marketplace to activate, unlock, upgrade, or extend usage of the application.

- Your application may not sell, link to, or otherwise promote mobile voice plans.
- Your application may not consist of, distribute, link to, or incentivize users to download or otherwise promote alternate marketplaces for applications and/or games.
- Your application must not jeopardize the security or functionality of Windows Phone devices or the Windows Phone Marketplace.
- If your application includes or displays advertising, the advertising must comply with the Microsoft Advertising Creative Acceptance Policy Guide (http://advertising.microsoft.com/creative-specs).
- If your application enables chat, instant messaging, or other person-to-person communication and allows the user to set up or create her account or ID from the mobile device, the application must include a mechanism to verify that the user creating the account or ID is at least 13 years old.
- If your application publishes a user's personal information from the mobile device to any service or other person, the application must implement "opt-in" consent. "Personal information" means all information or data associated with an identifiable user, including but not limited to the following, whether stored on the mobile device or on a Web-based server that is accessible from the mobile device:
  - Location information
  - Contacts
  - Photos
  - Phone number
  - SMS or other text communication
  - Browsing history
- To implement "opt-in" consent, the application must first describe how the personal information will be used or shared; obtain the user's express permission before publishing the information as described; and provide a mechanism through which the user can later opt out of having the information published.
- If your application allows users to purchase music content, it must include the Windows Phone music Marketplace (if available) as a

purchase option. If the application also allows music content to be purchased from any source other than the Windows Phone music Marketplace, the application must include its own playback functionality for that music content.

- If your application uses the Microsoft Push Notification Service (PNS), the application and the use of the PNS must comply with the following requirements.
  - The application must first describe the notifications to be provided and obtain the user's express permission (opt-in), and must provide a mechanism through which the user can opt out of receiving push notifications. All notifications provided using PNS must be consistent with the description provided to the user and must comply with all applicable Application Policies and 3.0 Content Policies.
  - The application and its use of the PNS must not excessively use network capacity or bandwidth of the PNS or otherwise unduly burden a Windows Phone or other Microsoft device or service with excessive push notifications, as determined by Microsoft in its reasonable discretion, and must not harm or interfere with any Microsoft networks or servers or any third-party servers or networks connected to the PNS.
- The PNS may not be used to send notifications that are mission-critical or otherwise could affect matters of life or death, including, without limitation, critical notifications related to a medical device or condition.

Content Policies

In addition to the basic application policies, Microsoft will also limit the types of applications based on the content of the application. This means several types of applications will not be allowed.

- Applications that promote illegal activities that are obscene or indecent as deemed under local laws.
- Applications that show or encourage harm to animals or persons in the real world.
- Applications that are defamatory, libelous, slanderous, or threatening.

- Applications that promote hate speech or are defamatory.
- Applications that could be used to sell (illegally or in excess) tobacco, drugs, weapons, or alcohol.
- Applications that allow the user to use a weapon in the real world (e.g., no remote hunting programs).
- Applications containing adult content, including nudity, sex, pornography, prostitution, or sexual fetishes or content that is sexual as it relates to children or animals.
- Applications containing realistic or gratuitous violence or gore. This also includes no content that shows rape (or suggestions of rape), molestation, instructions on injuring people in the real world, or glorification of genocide or torture.
- Applications containing excessive use of profanity.

The use of these policies is to promote a safe device for all ages. Since there currently is no way to control the use of the device by people of age, some of the content limitations are fairly restrictive. Microsoft seems committed to help developers by making sure everyone knows the extent of the policies and, when an application fails to be certified, to be very clear about how the application failed and including suggestions on how to change the application to allow it to pass.

Where Are We?

Windows Phone represents a platform, not just a device. To application developers, Windows Phone should represent an exciting new platform and software delivery mechanism. While developing for the phone is a very new experience, the basic tooling is not new. The fundamental underpinnings of Silverlight and XNA mean the platforms are mature and ready to develop for. This first chapter should excite you about the possibilities of creating great user experiences.

By combining a great device, a great platform, and the Marketplace, Microsoft has enabled you to be successful even while abiding by the rules in the Marketplace. For the most part, these rules are easy to live with. Are you ready to make money on the next exciting platform?

▪ 2 ▪
Writing Your First Phone Application

While the press might have you believe that becoming a phone-app millionaire is a common occurrence, it's actually pretty rare, but that doesn't mean you won't want to create applications for the phone. Hopefully the days of cheap and useless but popular phone apps are over, and we can start focusing on phone-app development as being a way to create great experiences for small and large audiences. Microsoft's vision of three screens is becoming a reality, as the phone is joining the desktop and the TV as another vehicle for you to create immersive experiences for users.

Although understanding Windows Phone capabilities and services is a good start, you are probably here to write applications. With that in mind, this chapter will walk you through setting up a machine for authoring your very first Windows Phone Silverlight application.

Preparing Your Machine

Before you can start writing applications for the phone, you must install the Windows Phone Developer Tools. Go to http://create.msdn.com to download the tools called Windows Phone SDK 7.1. This website is the starting point for downloading the tools as well as accessing the forums if you have further questions about creating applications.

Windows Phone Versioning Confusion

At the time of this writing there is a difference in how the phone and the underlying operating system are named. The phone itself is marketed as "Windows Phone 7.5" but the operating system is called "Windows Phone OS 7.1" and the development tools are called "Windows Phone SDK 7.1." This can be confusing, but if you remember the phone is "7.5" and all the software is "7.1" it can help.

To install the Windows Phone SDK 7.1 you must meet the minimum system requirements shown in Table 2.1.

Once you meet the requirements, you can run the vm_web.exe file that you downloaded from the website to install the Windows Phone SDK 7.1. The SDK installer includes Microsoft Visual Studio 2010 Express for Windows Phone, Microsoft Blend Express for Windows Phone (the Express version of Microsoft Expression Blend), and the Software Development Kit (SDK). Visual Studio Express is the coding environment for Windows Phone. Blend Express is the design tool for phone applications. And the SDK is a set of libraries for creating phone applications and an emulator for creating applications without a device.

TABLE 2.1 Windows Phone Developer Tools Requirements

Requirement	Description
Operating system	Windows 7, x86 or x64 (all but Starter Edition); or Windows Vista SP2, x86 or x64 (all but Starter Edition)
Memory	3GB RAM
Disk space	4GB free space
Graphics card	DirectX 10 capable card with a WDDM 1.1 driver

> **■ Tip**
>
> The Windows Phone SDK 7.1 does not work well in a virtual machine (e.g., Virtual PC, VMware, etc.) and is not officially supported. The emulator is a virtual machine of the phone, so running a virtual machine in a virtual machine tends to cause problems, especially slow performance.

Visual Studio is the primary tool for writing the code for your phone applications. Although the Windows Phone SDK 7.1 installs a version of Visual Studio 2010 Express specifically for phone development, if you already have Visual Studio 2010 installed on your machine the phone tools will also be integrated into this version of Visual Studio. The workflow for writing code in both versions of Visual Studio is the same. Although both versions offer the same features for developing applications for the phone, in my examples I will be using Visual Studio Express Edition for Windows Phone. In addition, I will be using Blend Express, not the full version of Blend (i.e., Expression Blend).

Creating a New Project

To begin creating your first Windows Phone application you will want to start in one of two tools: Visual Studio or Expression Blend. Visual Studio is where most developers start their projects, so we will begin there, but we will also discuss how you can use both applications for different parts of the development process.

Visual Studio

As noted earlier, when you install the Windows Phone SDK 7.1 you get a version of Visual Studio 2010 Express that is used to create Windows Phone applications only. When you launch Visual Studio 2010 Express you will see the main window of the application, as shown in Figure 2.1.

Click the New Project icon on the Start page and you will be prompted to start a new project. Visual Studio 2010 Express only supports creating applications for Window Phone. In the New Project dialog (see Figure 2.2)

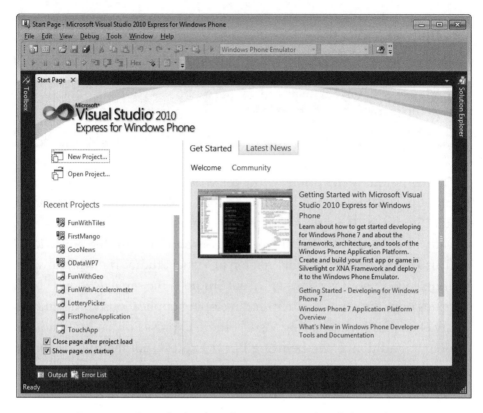

FIGURE 2.1 Microsoft Visual Studio 2010 Express for Windows Phone

you will notice that only Silverlight and XNA project templates are shown. For our first project we will start with a new Windows Phone Application template and name it "HelloWorldPhone".

When you click the OK button to create the project, Visual Studio will prompt you with a dialog where you can pick what version of the phone to target (version 7.0 or 7.1), as seen in Figure 2.3.

Once Visual Studio creates the new project, you can take a quick tour of the user interface (as shown in Figure 2.4). By default, Visual Studio shows two main panes for creating your application. The first pane (labeled #1 in the figure) is the main editor surface for your application. In this pane, every edited file will appear separated with tabs as shown. By default, the MainPage.xaml file is shown when you create a new Windows Phone application; this is the main design document for your new application.

FIGURE 2.2 **New Project dialog**

FIGURE 2.3 **Picking the phone version to target**

The second pane (#2 in the figure) is the Solution Explorer pane and it displays the contents of the new project.

Another common pane that you will use is the toolbar, and it is collapsed when you first use Visual Studio 2010 Express for Windows Phone. On the left side of the main window you will see a Toolbox tab that you can click to display the Toolbox, as shown in Figure 2.5.

Figure 2.4 The Visual Studio user interface

You may also want to click the pin icon to keep the toolbar shown at all times (as highlighted in Figure 2.5).

Before we look at how to create the application into something that is actually useful, let's see the application working in the device. You will notice that in the toolbar (not the Toolbox) of Visual Studio there is a bar for debugging. On that toolbar is a drop-down box for specifying what to do to debug your application. This drop down should already display the words "Windows Phone Emulator" as that is the default when the tools are installed (as shown in Figure 2.6).

At this point, if you press the F5 key (or click the triangular play button on the debugging toolbar), Visual Studio will build the application and start the emulator with our new application, as shown in Figure 2.7.

This emulator will be the primary way you will debug your applications while developing applications for Windows Phone. Our application

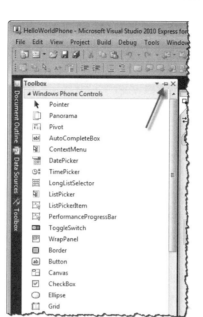

FIGURE 2.5 Enabling the toolbar

FIGURE 2.6 Using the emulator

FIGURE 2.7 The emulator

does not do anything, so you can go back to Visual Studio and click the square stop button on the debugging toolbar (or press Shift-F5) to end your debugging session. You should note that the emulator does not shut down. It is meant to stay running between debugging sessions.

XAML

In Silverlight, development is really split between the design and the code. The design is accomplished using a markup language called **XAML** (rhymes with camel). XAML (or eXtensible Application Markup Language) is an XML-based language for representing the look and feel of your applications. Since XAML is XML-based, the design consists of a hierarchy of elements that describe the design. At its most basic level, XAML can be used to represent the objects that describe the look and feel of an application.[1] These objects are represented by XML elements, like so:

```
<Rectangle />

<!-- or -->

<TextBox />
```

You can modify these XML elements by setting attributes to change the objects:

```
<Rectangle Fill="Blue" />

<!-- or -->

<TextBox Text="Hello World" />
```

Containers in XAML use XML nesting to imply ownership (a parent-child relationship):

```
<Grid>
  <Rectangle Fill="Blue" />
  <TextBox Text="Hello World" />
</Grid>
```

Using this simple XML-based syntax you can create complex, compelling designs for your phone applications. With this knowledge in hand, we

---

1. This is an oversimplification of what XAML is. Chapter 3, XAML Overview, will explain the nature of XAML in more detail.

can make subtle changes to the XAML supplied to us from the template. We could modify the XAML directly, but instead we will start by using the Visual Studio designer for the phone. In the main editor pane of Visual Studio the MainPage.xaml file is split between the designer and the text editor for the XAML. The left pane of the MainPage.xaml file is not just a preview but a fully usable editor. For example, if you click on the area containing the words "page name" on the design surface, it will select that element in the XAML, as shown in Figure 2.8.

Once you have that element selected in the designer, the properties for the element are shown in the Properties window (which shows up below the Solution Explorer). If the window is not visible, you can enable it in the View menu by selecting "Properties window" or by pressing the F4 key. This window is shown in Figure 2.9.

The Properties window consists of a number of small parts containing a variety of information, as shown in Figure 2.10.

The section near the top (#1 in Figure 2.10) shows the type of object you have selected (in this example, a `TextBlock`) and the name of the object, if any (here, "PageTitle"). This should help you ensure that you have selected the right object to edit its properties. The next section down (#2) contains a Search bar where you can search for properties by name, as well as buttons for sorting and grouping the properties. The third section (#3) is a list of the properties that you can edit.

▪ Note

You can also use the Properties window to edit events, but we will cover that in a later chapter.

FIGURE 2.8 Using the Visual Studio XAML design surface

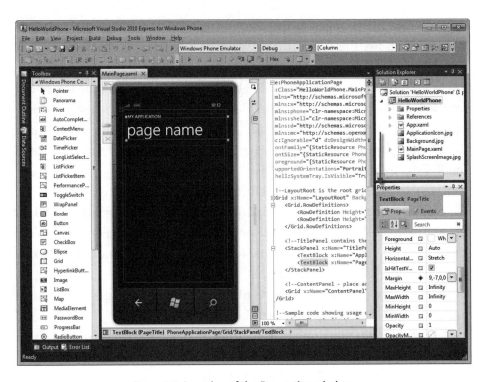

FIGURE 2.9 Location of the Properties window

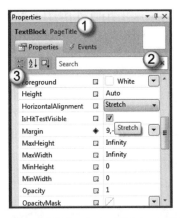

FIGURE 2.10 Contents of the Properties window

From the Properties window you can change the properties of the selected item. For example, to change the text that is in the TextBlock, you can simply type in a new value for the Text property. If you enter "hello world" in the Text property and press Return, the designer will change to display the new value. Changing this property actually changes the XAML in the MainPage.xaml file. The design surface is simply reacting to the change in the XAML. If you look at the XAML the change has been affected there as well, as shown in Figure 2.11.

You can edit the XAML directly as well if you prefer. If you click on the `TextBlock` above the `PageTitle` (the one labeled "ApplicationTitle"), you can edit the `Text` attribute directly. Try changing it to "MY FIRST WP7 APP" to see how it affects the designer and the Properties window:

```
...
<TextBlock x:Name="ApplicationTitle"
           Text="MY FIRST WP7 APP"
           Style="{StaticResource PhoneTextNormalStyle}" />
...
```

Depending on their comfort level, some developers will find it easier to use the Properties window while others will be more at ease editing the XAML directly. There is no wrong way to do this.

Although the Visual Studio XAML designer can create interesting designs, the real powerhouse tool for designers and developers is Blend. Let's use it to edit our design into something useful for our users.

FIGURE 2.11 The changed property

Designing with Blend

As noted earlier, in addition to offering an Express version of Visual Studio the Windows Phone SDK 7.1 includes an Express version of Expression Blend specifically for use in developing phone applications. You can launch Blend by looking for the shortcut key, or you can open it directly with Visual Studio. If you right-click on the MainPage.xaml file you will get a context menu like the one shown in Figure 2.12.

When you select Open in Expression Blend, Blend will open the same solution in the Expression Blend tool with the selected XAML file in the editor, as shown in Figure 2.13. You should save your project before going to Blend to make sure Blend loads any changes (Ctrl-Shift-S).

Although Expression Blend is thought of as purely a design tool, designers and developers alike can learn to become comfortable with it. And although Visual Studio and Expression Blend share some of the same features, both developers and designs will want to use Blend to build their designs. Some tasks are just simpler and faster to do in Blend. Chapter 5, Designing for the Phone, covers what tasks are better suited to Expression Blend.

Like Visual Studio, Blend consists of a number of panes that you will need to get familiar with.

FIGURE 2.12 Opening Blend directly in Visual Studio

FIGURE 2.13 The Blend user interface

> ■ **Note**
>
> Blend and Visual Studio both open entire solutions, not just files. This is a significant difference from typical design tools.

The first pane (labeled #1 in Figure 2.13) contains multiple tabs that give you access to several types of functionality. By default, the first tab (and the one in the foreground) is the Projects tab (though you could be showing a different tab by default). This tab shows the entire solution of projects. The format of this tab should look familiar; it's showing the same information as the Solution Explorer in Visual Studio. The next pane (#2) is the editor pane. This pane contains tabs for each opened file (only one at this point). MainPage.xaml should be the file currently shown in the editor. Note that the editor shows the page in the context of the phone so that you can better visualize the experience on the phone. On the right-hand side of the Blend interface is another set of tabs (#3) that contain information about selected items in the design surface. The selected tab should be the Properties tab.

This tab is similar to the Properties window in Visual Studio but is decidedly more designer-friendly. As you select items on the design surface, you'll be able to edit them in the Properties tab here. Finally, the Objects and Timeline pane (#4) shows the structure of your XAML as a hierarchy.

Let's make some changes with Blend. First (as shown in Figure 2.14) select the "hello world" text in the designer.

Once it's selected, you can see that eight squares surround the selection. These are the handles with which you can change the size or location of the `TextBlock`. While this object is selected, the Objects and Timeline pane shows the item selected in the hierarchy; as well, the item is shown in the Properties tab so that you can edit individual properties (as shown in Figure 2.15).

FIGURE **2.14 Selecting an object in Blend**

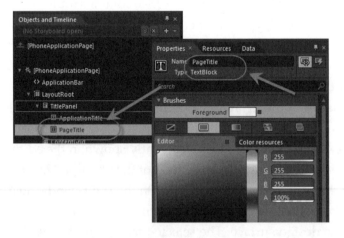

FIGURE **2.15 Selecting an object to edit in the Properties pane**

If you type "text" into the search bar of the Properties pane, the properties that have that substring in them will appear (to temporarily reduce the number of properties in the Properties pane). You can change the title by changing the `Text` property, as shown in Figure 2.16.

Once you're done changing the text, you may want to click the "X" in the Search bar to clear the search criteria. This will remove the search and show all the properties of the `TextBlock` again.

Selecting items and changing properties seems similar to what you can do in Visual Studio, but that's just where the design can start. Let's draw something. Start by selecting a container for the new drawing. In the Objects and Timeline pane, choose the `ContentPanel` item. This will show you that it is a container that occupies most of the space below our "hello world" text on the phone's surface.

We can draw a rectangle in that container by using the left-hand toolbar. On the toolbar you'll see a rectangle tool (as shown in Figure 2.17). Select the tool and draw a rectangle in the `ContentPanel` to create a new rectangle (also shown in Figure 2.17). If you then select the top arrow tool (or press the V key) you'll be able to modify the rectangle.

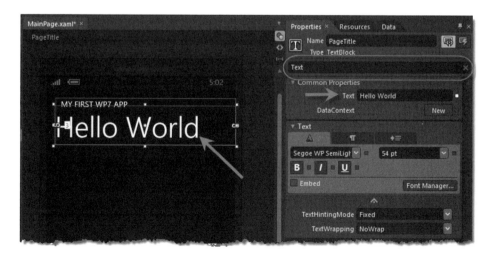

FIGURE 2.16 Updating a property in Blend

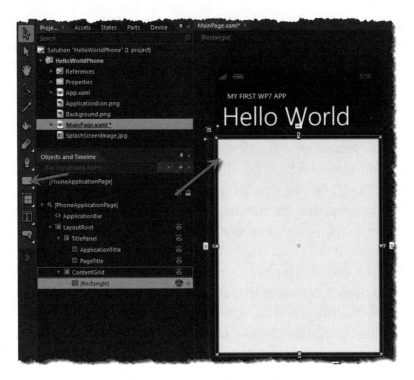

FIGURE 2.17 Drawing in a container

The rectangle you created has eight control points (the small squares at the corners and in the middle of each side). In addition, the rectangle has two small control points in the upper-left side (outside the surface area of the rectangle). These controls are used to round the corners of rectangles. Grab the top one with your mouse and change the corners to be rounded slightly, as shown in Figure 2.18.

Now that you have rounded the corners you can use the Properties pane to change the colors of the rectangle. In the Properties pane is a Brushes section showing how the different brushes for the rectangle are painted.

FIGURE 2.18 Rounding the corners

FIGURE 2.19 Editing brushes

The rectangle contains two brushes: a fill brush and a stroke brush. Selecting one of these brushes will allow you to use the lower part of the brush editor to change the look of that brush. Below the selection of brush names is a set of tabs for the different brush types, as shown in Figure 2.19.

The first four tabs indicate different options for brushes. These include no brush, solid color brush, gradient brush, and tile brush. Select the stroke brush, and then select the first tab to remove the stroke brush from the new rectangle. Now select the fill brush, and change the color of the brush by selecting a color within the editor, as shown in Figure 2.20.

Now let's put some text in the middle of our design to show some data. More specifically, let's put a `TextBlock` on our design. Go back to the toolbar and double-click the TextBlock tool (as shown in Figure 2.21). Although we drew our rectangle, another option is to double-click the toolbar, which will insert the selected item into the current container (in

FIGURE 2.20 Picking a color

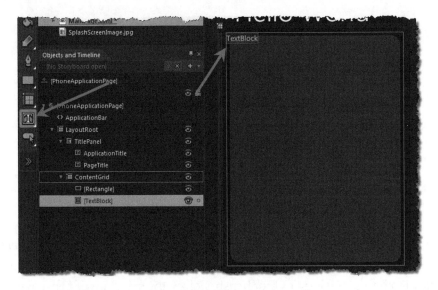

FIGURE 2.21 **Inserting a** `TextBlock`

this case, the `ContentPanel`). The inserted `TextBlock` is placed in the upper left of our `ContentPanel`, as also shown in Figure 2.21.

Once the new `TextBlock` is inserted, you can simply type to add some text. Type in "Status" just to have a placeholder for some text we will place later in this chapter. You should use the mouse to click on the Selection tool (the top arrow on the toolbar) so that you can edit the new `TextBlock`. You could use the mouse to place the `TextBlock` exactly where you like, but you could also use the Properties pane to align it. In the Properties pane, find the Layout section and select the horizontal center alignment and vertical bottom alignment, as shown in Figure 2.22. You might need to

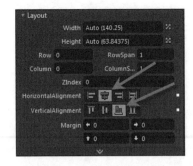

FIGURE 2.22 **Centering the** `TextBlock`

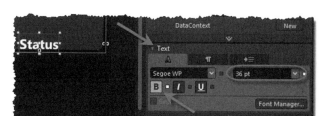

FIGURE 2.23 Changing the text properties

set your margins to zero as well to achieve the effect (as Blend may put a margin on your item depending on how you draw it).

Next you can edit the font and size of the TextBlock using the Text section of the Properties pane. You will likely need to scroll down to reach the Text section. From there you can change the font, font size, and text decoration (e.g., bold, italic, etc.). Change the font size to 36 points and make the font bold, as shown in Figure 2.23.

At this point our application does not do much, but hopefully you have gotten your first taste of the basics of using Blend for design. To get our first application to do something, we will want to hook up some of the elements with code. So we should close Blend and head back to Visual Studio.

When you exit Blend you will be prompted to save the project. Upon returning to Visual Studio your changes will be noticed by Visual Studio; allow Visual Studio to reload the changes.

> ### ■ Tip
>
> Blend is great at a variety of design tasks, such as creating animations, using behaviors to interact with user actions, and creating transitions. In subsequent chapters we will delve much further into using those parts of the tool.

Adding Code

This first Windows Phone application is not going to do much, but we should get started and make something happen with the phone. Since this is your first Windows Phone application, let's not pretend it is a desktop application but instead show off some of the touch capabilities.

First, if you look at the text of the XAML you should see that the first line of text shows the root element of the XAML to be a `PhoneApplicationPage`. This is the basic class from which each page you create will derive. The `x:Class` declaration is the name of the class that represents the class. If you open the code file, you will see this code was created for you.

```
<phone:PhoneApplicationPage x:Class="HelloWorldPhone.MainPage"
   . . .
```

> ■ **Note**
>
> The "phone" alias is an XML alias to a known namespace. If you're not familiar with how XML namespaces work we will cover it in more detail in Chapter 3.

You will want to open the code file for the XAML file. You can do this by right-clicking the XAML page and picking View Code or you can simply press F7 to open the code file. The initial code file is pretty simple, but you should see what the basics are. The namespace and class name match the `x:Class` definition we see in the XAML. This is how the two files are related to each other. If you change one, you will need to change the other. You should also note that the base class for the `MainPage` class is the same as the root element of the XAML. They are all related to each other.

```
using System;
using System.Collections.Generic;
using System.Linq;
using System.Net;
using System.Windows;
using System.Windows.Controls;
using System.Windows.Documents;
using System.Windows.Input;
using System.Windows.Media;
using System.Windows.Media.Animation;
using System.Windows.Shapes;
using Microsoft.Phone.Controls;

namespace HelloWorldPhone
{
    public partial class MainPage : PhoneApplicationPage
    {
```

```
    // Constructor
    public MainPage()
    {
      InitializeComponent();
    }
  }
}
```

These two files (the .xaml and the code files) are closely tied to each other. In fact, you can see that if you find an element in the XAML that has a name it will be available in the code file. If you switch back to the .xaml file, click on the TextBlock that you created in Blend. You will notice in the Properties window that it does not have a name (as shown in Figure 2.24).

If you click where it says "<no name>" you can enter a name. Name the TextBlock "theStatus". If you then switch over to the code file, you will be able to use that name as a member of the class:

```
...
public partial class MainPage : PhoneApplicationPage
{
  // Constructor
  public MainPage()
  {
    InitializeComponent();

    theStatus.Text = "Hello from Code";
  }
}
...
```

At this point, if you run the application (pressing F5 will do this) you will see that this line of code is being executed as the theStatus Text-Block is changed to show the new text (as seen in Figure 2.25).

There is an important fact you should derive from knowing that named elements in the XAML become part of the class: The job of the XAML is to

FIGURE 2.24 Naming an element in the Properties window

Figure 2.25 Running the application

build an object graph. The hierarchy of the XAML is just about creating the hierarchy of objects. At runtime, you can modify these objects in whatever way you want.

When you stop your application the emulator will continue to run. You can leave the emulator running across multiple invocations of your application. You should not close the emulator after debugging your application.

Working with Events

Since you are building a phone application, let's show how basic events work. You can wire up events just as easily using standard language (e.g., C#) semantics.[2] For example, you could handle the MouseLeftButtonUp event on theStatus to run code when the text is tapped:

```
...
public partial class MainPage : PhoneApplicationPage
{
  // Constructor
  public MainPage()
  {
    InitializeComponent();
```

---

2. For Visual Basic, you would just use the handles keyword instead of the C# event handler syntax.

```
    theStatus.Text = "Hello from Code";

    theStatus.MouseLeftButtonUp +=
        new MouseButtonEventHandler(theStatus_MouseLeftButtonUp);
  }

  void theStatus_MouseLeftButtonUp(object sender,
                                   MouseButtonEventArgs e)
  {
    theStatus.Text = "Status was Tapped";
  }
}
...
```

When you tap on theStatus the MouseLeftButtonUp event will be fired (which is what causes the code in the event handler to be called). All events work in this simple fashion, but the number and type of events in Silverlight for Windows Phone vary widely.

Debugging in the Emulator

If clicking the user interface was not working the way we would like, it might help if we could stop the operation during an event to see what was happening during execution. We can do this by debugging our operation. We can use the debugger to set breakpoints and break in code while using the emulator. Place the text cursor inside the event handler and press F9 to create a breakpoint. When you run the application (again, press F5) you can see that when you click on the theStatus TextBlock the debugger stops inside the event handler. You can hover your mouse over specific code elements (e.g., theStatus.Text) to see the value in a pop up (as shown in Figure 2.26).

Pressing the F5 key while stopped at a breakpoint will cause the application to continue running. There are other ways to walk through the code, but for now that should be sufficient to get you started. Using the emulator is the most common way you will develop your applications, but there are some interactions that are difficult to do with the emulator (e.g., multi-touch, using phone sensors, etc.) for which debugging directly on a device would be very useful. Luckily, debugging on the device is supported and works pretty easily.

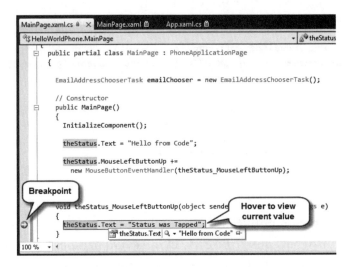

FIGURE 2.26 **Using the Visual Studio debugger**

Debugging with a Device

If you have a phone with which you want to do your development, you will want to be able to deploy and debug directly on the phone itself. First you need to connect your phone to your development machine. All communication with the phone is routed through the Zune software. By connecting your device to your computer, the Zune software should start automatically. If it does not start up, you can run it manually. Although the device will continue to operate normally when connected to Zune, several functions are disabled to allow Zune to synchronize media (music, photos, and videos) to the device. Consequently, these functions are using your media on the phone. Once connected, the device looks like it would normally (as seen in Figure 2.27).

All the communication that you will do to your phone (e.g., debugging, deploying, and registration) is done while Zune is running. Once you've connected your phone to your computer and run Zune, you should be able to see the phone attached, as shown in Figure 2.28.

Now that your device is connected, you can use it to sync your music, videos, and photos to the phone. However, before you can use a phone as a development device, you will need to register the phone for development.

FIGURE 2.27 Connected device

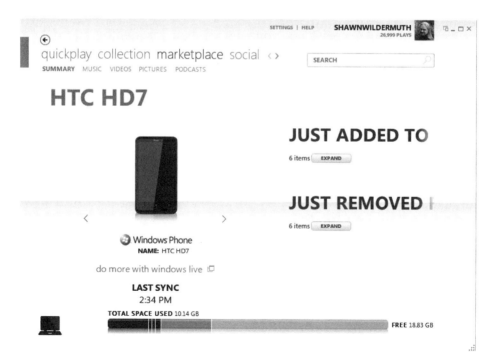

FIGURE 2.28 Your phone connected to the Zune software

This lifts the requirements that applications be signed by Microsoft, and allows you to deploy your applications directly to the phone so that you can debug applications.

Before you can enable your phone as a developer phone, you will need to have an account at the Windows Phone developer portal (http://developer .windowsphone.com). Once you have done that, you can enable your phone to be used for development. To do this you will need the Windows Phone Developer Registration tool, which is installed when you install the Windows Phone SDK 7.1. When you run this application it will ask you for your Windows Live ID that you used to register with the developer portal, as shown in Figure 2.29.

If your phone is successfully attached to your computer, the Status area will tell you that it is ready to register your device for development. At this point, just click the Register button to register with the developer portal. Once it registers the phone, it changes the status to show you that the phone is ready (as shown in Figure 2.30).

Now that you've registered your device, you can deploy and debug your applications using Visual Studio. The key to using the device instead

FIGURE 2.29 **Registering your device**

Figure 2.30 Successfully registered developer phone

of the emulator is to change the deployment using the drop-down list of deployment options. There are only two:

- Windows Phone Emulator
- Windows Phone Device

The drop down is located in the toolbar of Visual Studio, as shown in Figure 2.31.

Once you change the deployment target you can debug just like you did with the emulator. When you run the application, it will deploy your application to the device and run it so that you can debug it in the same way as you did with the emulator. Figure 2.32 shows the application running on a device.

Figure 2.31 Changing the deployment to use a development phone

FIGURE 2.32 **Running on a device**

Using Touch

Even though the touch interactions do fire mouse events, there are other events that allow you to design your application for touch. Since touch is so important to how applications on the phone work, this first application should give you a taste of that experience. To show touch working, let's add an ellipse to the application that the user can move around by dragging it with her finger. To get started, you should open the MainPage.xaml file and add a new ellipse in the center of the page. To do this find the TextBlock called theStatus and place a new Ellipse element after it, like so:

```
...
    <Grid x:Name="ContentGrid"
          Grid.Row="1">
      <Rectangle Fill="#FF7E0505"
                 Margin="8"
                 RadiusY="24"
                 RadiusX="24" />
      <TextBlock HorizontalAlignment="Center"
                 TextWrapping="Wrap"
                 Text="Status"
                 VerticalAlignment="Bottom"
                 FontSize="48"
                 FontWeight="Bold"
                 Name="theStatus" />
```

```
          <Ellipse x:Name="theEllipse"
                   Fill="White"
                   Width="200"
                   Height="200">
          </Ellipse>
      </Grid>
  ...
```

We want to be able to move the ellipse (named `theEllipse`) as the user drags it. To allow us to do this we will need to use something called a **transform.** In Silverlight, a transform is used to change the way an object is rendered without having to change properties of the ellipse. While we could change the margins and/or alignments to move it around the screen, using a transform is much simpler. You should use a `TranslateTransform` to allow this movement. A `TranslateTransform` provides X and Y properties, which specify where to draw the element (as a delta between where it originally exists and where you want it). You can specify this transform by setting the `RenderTransform` property with a `TranslateTransform` (naming it in the process):

```
  ...
  <Ellipse x:Name="theEllipse"
           Fill="White"
           Width="200"
           Height="200">
    <Ellipse.RenderTransform>
      <TranslateTransform x:Name="theMover" />
    </Ellipse.RenderTransform>
  </Ellipse>
  ...
```

Now that we have a way to move our ellipse around the page, let's look at dealing with touch. In Silverlight, there are two specific types of touch interactions that are meant to allow the user to change on-screen objects. These are when the user drags her finger on the screen and when she uses a pinch move to resize objects. These types of interactions are called **manipulations.** Silverlight has three events to allow you to use this touch information:

- `ManipulationStarted`
- `ManipulationDelta`
- `ManipulationCompleted`

These events let you get information about the manipulation as it happens. For example, let's handle the `ManipulationDelta` event to get information about when the user drags on the screen. This event is called as the manipulation happens, and it includes information about the difference between the start of the manipulation and the current state (e.g., how far the user has dragged her finger):

```
...
public partial class MainPage : PhoneApplicationPage
{
  // Constructor
  public MainPage()
  {
    InitializeComponent();

    theStatus.Text = "Hello from Code";

    theStatus.MouseLeftButtonUp +=
      new MouseButtonEventHandler(theStatus_MouseLeftButtonUp);

    theEllipse.ManipulationDelta +=
      new EventHandler<ManipulationDeltaEventArgs>(
                            theEllipse_ManipulationDelta);
  }

  void theEllipse_ManipulationDelta(object sender,
                                    ManipulationDeltaEventArgs e)
  {
    // As a manipulation is executed (drag or resize), this is called
    theMover.X = e.CumulativeManipulation.Translation.X;
    theMover.Y = e.CumulativeManipulation.Translation.Y;
  }

  ...
}
...
```

The event is fired while the user either pinches or drags within the `theEllipse` element. In this case the code is only concerned with the dragging. In the event handler for `ManipulationDelta`, the `ManipulationDeltaEventArgs` object contains information about the extent of the manipulation. In the `CumulativeManipulation` property of the event args, there is a property called `Translation`, which contains the extent of the drag operation (the complete delta). We are just changing

FIGURE 2.33 Dragging the ellipse

theMover's properties to match the manipulation. This means we can now drag the theEllipse element around and see it change position under our dragging, as seen in Figure 2.33.

Working with the Phone

This first application is a program that can be pretty self-sufficient, but not all applications are like that. Most applications will want to interact with the phone's operating system to work with other parts of the phone. From within your application you may want to make a phone call, interact with the user's contacts, take pictures, and so on. The Windows Phone SDK 7.1 calls these types of interactions **tasks.** Tasks let you leave an application (and optionally return) to perform a number of phone-specific tasks. Here is a list of some of the most common tasks:

- CameraCaptureTask
- EmailAddressChooserTask
- EmailComposeTask

- PhoneCallTask
- SearchTask
- WebBrowserTask

These tasks allow you to launch a task for the user to perform. In some of these tasks (e.g., CameraCaptureTask, EmailAddressChooser-Task), once the task is complete the user expects to return to your application; while in others (e.g., SearchTask) the user may be navigating to a new activity (and may come back via the Back key, but may not).

Let's start with a simple task, the SearchTask. Add a using statement to the top of the code file for Microsoft.Phone.Tasks to make sure the SearchTask class is available to our code file. Next, create an event handler for the MouseLeftButtonUp event on theEllipse. Then, inside the handler for the MouseLeftButtonUp event, you can create an instance of the SearchTask, set the search criteria, and call Show to launch the task:

```
...
using Microsoft.Phone.Tasks;
...
public partial class MainPage : PhoneApplicationPage
{
  // Constructor
  public MainPage()
  {
    ...

    theEllipse.MouseLeftButtonUp += new
      MouseButtonEventHandler(theEllipse_MouseLeftButtonUp);

  }

  void theEllipse_MouseLeftButtonUp(object sender,
                                    MouseButtonEventArgs e)
  {
    SearchTask task = new SearchTask();
    task.SearchQuery = "Windows Phone";
    task.Show();
  }

  ...
}
```

Figure 2.34 The `SearchTask` in action

If you run your application, you'll see that when you tap on the `theEllipse` element it will launch the phone's Search function using the search query you supplied (as shown in Figure 2.34). The results you retrieve for the search query may vary as it is using the live version of Bing for search.

While this sort of simple task is useful, the more interesting story is being able to call tasks that return to your application. For example, let's pick an email address from the phone and show it in our application. The big challenge here is that when we launch our application, we may get tombstoned (or deactivated). Remember that, on the phone, only one application can be running at a time. In order to have our task wired up when our application is activated (remember, it can be deactivated or even unloaded if necessary), we have to have our task at the page or application level and wired up during construction. So, in our page, create a class-level field and wire up the `Completed` event at the end of the constructor for it, like so:

```
public partial class MainPage : PhoneApplicationPage
{

  EmailAddressChooserTask emailChooser =
    new EmailAddressChooserTask();
```

```
// Constructor
public MainPage()
{
  ...

   emailChooser.Completed += new
     EventHandler<EmailResult>(emailChooser_Completed);
}


  ...
}
```

In the event handler, we can simply show the email chosen using the MessageBox API:

```
...
void emailChooser_Completed(object sender, EmailResult e)
{
   MessageBox.Show(e.Email);
}
...
```

Now we need to call the task. To do this, let's highjack the event that gets called when the `theEllipse` element is tapped. Just comment out the old `SearchTask` code and add a call to the `emailChooser`'s `Show` method, like so:

```
...
void theEllipse_MouseLeftButtonUp(object sender,
                                    MouseButtonEventArgs e)
{
   //SearchTask task = new SearchTask();
   //task.SearchQuery = "Windows Phone";
   //task.Show();

   // Get an e-mail from the user's Contacts
   emailChooser.Show();
}
...
```

Once you run the application, a list of contacts will be shown and you will be able to pick a contact (and an address, if there is more than one), as shown in Figure 2.35. The emulator comes prepopulated with several fake contacts to test with.

Once the user picks the contact, the phone returns to your application. You will be returned to your application (and debugging will continue).

FIGURE 2.35 Choosing a contact to retrieve an email address via the
`EmailAddressChooserTask`

FIGURE 2.36 Showing the selected email in a `MessageBox`

The event handler should be called when it is returned to the application, as shown in Figure 2.36.

Where Are We?

You have created your first Windows Phone application using Silverlight. Although this example has very little real functionality, you should have a feel for the environment. Taking this first stab at an application and applying it to your needs is probably not nearly enough. That's why, in subsequent chapters, we will dive deep into the different aspects of the phone application ecosystem. You will learn how to build rich applications using Silverlight, services, and the phone itself.

3

XAML Overview

While phones used to be very utilitarian (e.g., wow, I can make a phone call) things have changed dramatically over the past few years. Therefore, although you could try to create applications that are focused on functionality, they probably would not attract a large audience unless they had a great user interface. That is what draws people to an application. That means it's your job to figure out what people want and what makes an application easy to use and learn. Luckily Silverlight comes with a way to design interfaces with a lot of control over the look and feel of the application. Creating dynamic applications that will wow your users is easier than ever. This chapter will show you how.

What Is XAML?

What is eXtensible Application Markup Language (or XAML)? In Silverlight XAML is used to design the user interfaces (both the look and feel of applications), but it does quite a lot more. The main concept to learn here is that XAML can be thought of as a serialization format that works well with tools. It allows us to declare the structure of a user interface. Declaring the interface in this way makes it easy for tools to create the user interfaces and have applications consume the files at runtime.

What do I mean by a serialization format? XAML is quite simple; let's take a very basic piece of XAML:

```
<UserControl x:Class="WinningTheLottery.Sample"
    xmlns="http://schemas.microsoft.com/winfx/2006/xaml/presentation"
    xmlns:x="http://schemas.microsoft.com/winfx/2006/xaml">
  <Grid>
    <TextBlock Text="Hello" />
    <Rectangle Width="100"
               Height="100"
               Fill="Blue" />
  </Grid>
</UserControl>
```

XAML is an XML file that obeys basic XML rules (e.g., single top-level container, case sensitivity). In this file we are declaring a `UserControl` root that contains a `Grid` element that contains two elements (a `TextBlock` and a `Rectangle`). This is the basic hierarchy of this simple user interface. When parsed, this XAML document is used to create that same hierarchy in memory. Literally, the name of the element ties itself to the name of a class. So when the XAML is parsed, the `UserControl` element informs the system to create a new `UserControl` instance. To be used here all the classes must allow for empty constructors (in the .NET sense) so that the `UserControl` class can be created. After it creates the `UserControl` itself, it looks at its subelements (the `Grid`) and creates that element as a child inside the `UserControl`. Finally, it creates the `TextBlock` and `Rectangle` and places them as children inside the `Grid`. When the `TextBlock` is created, it sees the attribute (`Text`) and calls the property setter of the new `TextBlock` with the contents of the attribute. It does this with the multiple attributes of the `Rectangle` as well. In this way it uses the XAML to build an in-memory object graph that follows the same structure as the XAML. Understanding that the XAML you are using is the basis for your runtime design is very important in understanding how XAML works.

Silverlight 5 Developers

For those of you who are using Silverlight 5 for the desktop, the Silverlight version for the phone is based on Silverlight 4. This means it is missing features you might already be using, including implicit data templates and relative bindings, among others.

XAML Object Properties

Most objects' properties you will set in XAML are simple and string-based:

```
<Rectangle Fill="Blue" />
```

Not all properties can be set using the simple, string-based syntax, however. Under the covers, many properties (during XAML parsing) are attempting to convert a string attribute to a property value (using a special type of class called a `TypeConverter`). For example, `Fill` is a property that accepts a `Brush` value, not a color. When `Fill="Blue"` is parsed as XAML, a conversion is done between the string (i.e., "Blue") and a brush called a `SolidColorBrush`. For a more complex value type (like a `Brush`), there is a verbose syntax for setting property values:

```
<Rectangle>
  <Rectangle.Fill>
    <SolidColorBrush Color="Blue" />
  </Rectangle.Fill>
</Rectangle>
```

This verbose syntax is identical at runtime to the earlier example. By adding an element inside the `Rectangle` whose name is the name of the object, a dot, and the name of the property (e.g., `Rectangle.Fill`), we can define the value for the property using XAML instead of being stuck using just strings. Since not all complex property values can be defined in such a way that a conversion can be made, this syntax allows for property values to be set when the value is a complex type that would be difficult or impossible to describe in a single string. For example, let's replace the `SolidColorBrush` with a `LinearGradientBrush`:

```
<Rectangle>
  <Rectangle.Fill>
    <LinearGradientBrush>
      <GradientStop Color="Blue" Offset="0" />
      <GradientStop Color="White" Offset="0.5" />
      <GradientStop Color="Blue" Offset="1" />
    </LinearGradientBrush>
  </Rectangle.Fill>
</Rectangle>
```

In this example, we can see that defining a fill by specifying the colors and offsets not only would be difficult in a simple string, but would make the XAML even harder to read. In this way, XAML allows you to set very complex properties without having to invent conversions. You will see how this is used in many different places in XAML as we continue.

Understanding XAML Namespaces

Inside the root element are two namespaces. Namespaces in XAML are XML namespaces.[1] The default namespace (xmlns) declares that this is a XAML document. The second namespace (xmlns:x) brings in several elements and attribute types that are all prefixed with the x alias. So when you see x:Class, that is a convention that is defined in the second namespace.

You can think of the namespace aliasing as similar to namespaces in .NET. When you add a namespace, it brings in those new types of things that can be described in XAML. Unlike .NET, though, you have to use an alias (since all other namespaces are not the "default" namespace) and then use that alias everywhere you want to refer to information from that namespace. In the x alias's case, the alias here as "x" is just a convention that XAML tends to use. In the XML namespace sense, you can change the alias to whatever you want, but you would have to change it everywhere it's referenced as well. For example:

```
<UserControl foo:Class="WinningTheLottery.Sample"
    xmlns="..."
    xmlns:foo="...">
```

The alias is just that: an alias so that the parser can determine which of the namespaces your element or attribute is from. When we changed the name of the alias, it is what you would use to alias that namespace in the rest of the document.

While these namespaces represent the basic XAML namespaces, you can extend the XAML by using namespaces to bring in arbitrary .NET types as well. If you define a namespace that points at a .NET namespace and assembly, those types will also be available in the XAML:

```
<UserControl x:Class="WinningTheLottery.Sample"
    xmlns="http://schemas.microsoft.com/winfx/2006/xaml/presentation"
    xmlns:x="http://schemas.microsoft.com/winfx/2006/xaml"
```

---

1. http://shawnw.me/pFshpG

```
    xmlns:sys="clr-namespace:System;assembly=mscorlib">
    <Grid>
      <TextBlock>
        <TextBlock.Text>
          <sys:String>Hello</sys:String>
        </TextBlock.Text>
      </TextBlock>
    </Grid>
</UserControl>
```

In this example, the XAML "imports" the `System` namespace that exists inside the `mscorlib.dll` assembly. Once that .NET namespace is imported, all the types in that namespace are creatable in the XAML. Any type that is created in XAML must conform to the following rules:

- Has an empty, public constructor
- Has public properties

If any .NET objects follow these rules, they are creatable in XAML (and therefore can be part of your initial object graph). You will see how this is used as we continue in this chapter.

Naming in XAML

Unlike other platforms, XAML does not require that every object in the XAML be specifically named. In fact, it is probably a bad idea to name every object in the XAML. Naming objects in the XAML becomes important once you need to refer to an object by name (e.g., from code or via data binding). Naming objects in XAML takes the form of an attribute that can be applied to most XAML elements: `x:Name`. For example:

```
<UserControl x:Class="WinningTheLottery.Sample"
    xmlns="http://schemas.microsoft.com/winfx/2006/xaml/presentation"
    xmlns:x="http://schemas.microsoft.com/winfx/2006/xaml">
  <Grid x:Name="LayoutRoot">
    <TextBlock Text="Hello"
               x:Name="theTextBlock" />
    <Rectangle Width="100"
               Height="100"
               Fill="Blue"
               x:Name="theRectangle" />
  </Grid>
</UserControl>
```

You will notice that the naming attribute starts with the x: prefix (or alias). As was explained in the namespaces discussion, this means that attribute is available through the x namespace that is included on the top of every XAML document (by default). Once these objects are named, they will be available to other XAML elements by name or via code (both of which you will see in this chapter). The names used here must be unique. Each name can occur only once within a single XAML document. This simplifies the naming strategy but also means there is no sense of naming scope (like HTML has).

Visual Containers

If you consider the examples that have been shown, you may have missed the importance of XAML containers. The most obvious of these can be seen in the Grid element:

```
<UserControl x:Class="WinningTheLottery.Sample"
    xmlns="http://schemas.microsoft.com/winfx/2006/xaml/presentation"
    xmlns:x="http://schemas.microsoft.com/winfx/2006/xaml">
  <Grid>
    <TextBlock Text="Hello" />
  </Grid>
</UserControl>
```

The purpose of these containers is to allow other elements to be laid out in particular ways on the visual surface of Silverlight. The containers themselves typically don't have any user interface but simply are used to determine how different XAML elements are arranged on the screen. A number of layout containers are important to designing in XAML. Each of these can contain one or more child elements and lay them out in specific ways. You can see the common visual containers in Table 3.1.

These containers are important as they are used to determine how your elements are laid out. The most important of these is the Grid container and it will be the one you use most often. The Grid is a container that supports dynamic, table-like layout using rows and columns. To define rows and columns, you set the Grid's ColumnDefinitions and/or RowDefinitions properties. These properties take one or more ColumnDefinition or Row-Definition elements, as shown in the following code:

TABLE 3.1 Visual Containers

Layout Container	Description	Supports Multiple Children?
Grid	Table-like layout of columns and rows; good for alignment or margin-oriented design	Yes
StackPanel	Horizontal or vertical stacking of individual elements	Yes
Canvas	Position-based layout (via top and left positions)	Yes
ScrollViewer	Virtual container that can be larger than the contents to allow users to scroll through the container	No
Border	To create a simple border around a single element	No

```xml
<UserControl x:Class="WinningTheLottery.Sample"
    xmlns="http://schemas.microsoft.com/winfx/2006/xaml/presentation"
    xmlns:x="http://schemas.microsoft.com/winfx/2006/xaml">
  <Grid>
    <Grid.ColumnDefinitions>
      <ColumnDefinition />
      <ColumnDefinition />
    </Grid.ColumnDefinitions>
    <TextBlock Text="Hello" />
  </Grid>
</UserControl>
```

You create new columns and rows using the ColumnDefinitions and RowDefinitions properties (as shown). This allows you to specify that individual elements are in a particular row or column using the Grid.Column or Grid.Row attached properties (see the sidebar, What Are Attached Properties?):

```xml
<UserControl x:Class="WinningTheLottery.Sample"
    xmlns="http://schemas.microsoft.com/winfx/2006/xaml/presentation"
    xmlns:x="http://schemas.microsoft.com/winfx/2006/xaml">
  <Grid>
    <Grid.ColumnDefinitions>
      <ColumnDefinition />
      <ColumnDefinition />
```

```
    </Grid.ColumnDefinitions>
    <TextBlock Text="Hello"
              Grid.Column="1" /> <!-- The Second Column -->
  </Grid>
</UserControl>
```

By using the attached property, the TextBlock is indicating that the TextBlock belongs in the second column (note that row and column numbers are zero-indexed). In this way, the Grid is creating columns or rows proactively by specifying the number of rows or columns up front. At first blush it may seem verbose to create row and/or column definitions this way, but it's important as the definitions contain other important information that can be set.

What Are Attached Properties?

Some properties are not relevant until they exist in some specific scope. For example, when an object is inside a Grid, being able to tell the XAML what column or row you are in becomes critical. But that same element inside a StackPanel has no notion of a row or column. Attached properties are specific types of properties that are only valid in certain cases. Attached properties are defined by the name of the owning type and the name of the attached property (e.g., Grid.Row). The information in attached properties is available to the class that exposes them as that is where they are typically used. Though in XAML it is common for these attached properties to be used in this way, attached properties are really for properties that are global in scope. So you can define a property that could be applied to any XAML object. Containers such as the Grid and the Canvas expose attached properties to explicitly let them handle layout and will probably be the first real use of attached properties for most developers who are new to XAML.

For example, in the Grid class, as the Grid object lays out the elements inside it, it will query for the attached property to determine which row and/or column to place an element. Literally the properties are attached at runtime, so the underlying element does not need to have unneeded properties (such as Row and Column).

TABLE **3.2** `Grid` **Row and Column Sizing**

Type	Description	Example
Auto	Sizes row or column based on the contents. The size will be determined by the largest object in a respective row or column.	`<RowDefinition Height="Auto" />`
Pixel	Sets row or column to a specific size, in pixels. Larger objects will be clipped.	`<RowDefinition Height="100" />`

Star	Proportionally sizes rows or columns based on the weighted value.	`<RowDefinition Height="*" />`
		`<RowDefinition Height="25*" />`
		`<RowDefinition Height="0.147*" />`

When creating rows and columns, you can define the height or width (respectively) in three ways, as shown in Table 3.2.

Auto and pixel sizing are pretty self-explanatory, but star sizing requires some explanation. Star sizing proportionally sizes rows or columns based on the values of the height or width. For example:

```
<UserControl x:Class="WinningTheLottery.Sample"
    xmlns="http://schemas.microsoft.com/winfx/2006/xaml/presentation"
    xmlns:x="http://schemas.microsoft.com/winfx/2006/xaml">
  <Grid>
    <Grid.ColumnDefinitions>
      <ColumnDefinition Width="33*" />
      <ColumnDefinition Width="66*" />
    </Grid.ColumnDefinitions>
  </Grid>
</UserControl>
```

The width values are used as weighted proportions of the whole size. While this looks similar to percentages (like you may be used to in Web applications), the numbers are not part of an arbitrary 100% scale. For example, changing the values to `"1*"` and `"2*"` will yield the same 2-to-1 ratio as `"33*"` and `"66*"`. In the case of using a star alone (e.g., "*"), it is equivalent to `"1*"`. By using `Grid` elements with a mix of auto, pixel, and star sizing, you can create elastic layouts (using star sizing for the flexible sized elements and pixel/auto sizing for the more static parts of the design).

You have already seen that you can use attached properties to set the row and/or column of a specific element inside the `Grid`. The `Grid` class also supports the ability to specify `RowSpan` and `ColumnSpan` to signify that a particular element should span more than one row and/or column. This will give you extra flexibility to create your table-based designs, like so:

```xml
<Grid>
  <Grid.ColumnDefinitions>
    <ColumnDefinition Width="*" />
    <ColumnDefinition Width="*" />
    <ColumnDefinition Width="*" />
  </Grid.ColumnDefinitions>
  <Grid.RowDefinitions>
    <RowDefinition Height="*" />
    <RowDefinition Height="*" />
  </Grid.RowDefinitions>
  <TextBlock Text="1" />
  <TextBlock Text="1"
             Grid.Row="1" />
  <TextBlock Text="1"
             Grid.Column="1" />
  <TextBlock Text="Across All 3 Columns"
             Grid.ColumnSpan="3" />
  <TextBlock Text="Across Both Rows"
             Grid.RowSpan="2" />
</Grid>
```

Although you may use the other layout containers in certain cases, you should become comfortable with the `Grid` as it is the container you will use most often.

Visual Grammar

Silverlight gives you the ability to draw shapes and colors on the surface of the phone itself. While you might not imagine doing much actual drawing, it is important for you to understand how creating a design with the drawing primitives is important to the overall XAML story. As you start to use controls, you will learn that those controls are made up of more primitive elements, and when you want to change the way controls and other elements look, you will have to understand the drawing stack.

Shapes

The most basic drawing element is a `Shape`. The `Shape` element is a base class for a small number of shapes that are used for basic drawing. The basic shapes are listed below:

- Line
- Rectangle
- Ellipse
- Polygon
- Polyline
- Path

Each shape has basic attributes, such as `Height`, `Width`, `Fill`, and `Stroke`:

```
<Grid>
   <Rectangle Width="100"
              Height="100"
              Fill="Blue" />
   <Ellipse Width="200"
            Height="50"
            Stroke="Black" />
</Grid>
```

If you cannot compose the kind of shape you need with the first five shapes in the previous list, everything falls down to the `Path` shape. A `Path` is a powerful shape that can give you full power to design arbitrary shapes. The `Path` shape allows you to create open, closed, or compound shapes. The `Path` shape has a property called `Data` that specifies the elements of the shape. For example, you can specify a `Path` with an object graph, like so:

```
<Path Stroke="Black">
  <Path.Data>
    <PathGeometry>
      <PathFigure StartPoint="0,50">
        <BezierSegment Point1="50,0"
                       Point2="50,100"
                       Point3="100,50"/>
      </PathFigure>
```

```
      </PathGeometry>
    </Path.Data>
  </Path>
```

By setting the `Data` attribute to a `PathGeometry` element that specifies a `Path` that contains a `BezierSegment` (from 0,50 to 100,50 with control points of 50,0 and 50,100 as the curves), you can draw a curved line, as shown in Figure 3.1.

The `Data` property contains a shorthand notation to simplify and shorten the size of the XAML. This shorthand is the same type of information, but stored in a single string. For example, the curve in Figure 3.1 can be simplified to:

```
<Path Stroke="Black"
        Data="M 0,50 C 50,0 50,100 100,50" />
```

This has the same information in it but in a shortened form: Move to the 0,50 position and do a Bezier curve using these three points. Usually `Paths` are created with tools (e.g., Expression Blend) as the process can get terse, but the process does allow for very complex paths.

Brushes

In many of the examples so far, you have seen color names (e.g., black, red) used in XAML to indicate what color an object is displayed with. In fact, those colors were a shortcut to creating an object called a brush. Brushes are always used to paint surfaces (e.g., using fill, stroke, and background brushes). Several types of brushes are available to you, as shown in Table 3.3.

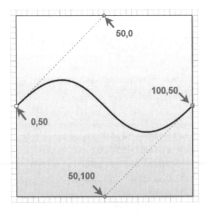

FIGURE 3.1 Path explained

TABLE 3.3 Brush Types

Type	Description	Example
SolidColorBrush	Paints a solid color	`<Ellipse Fill="Blue" />`

LinearGradientBrush	Paints a gradient along a line	`<Ellipse>`
		`  <Ellipse.Fill>`
		`    <LinearGradientBrush>`
		`      <GradientStop Color="Blue"`
		`                    Offset="0" />`
		`      <GradientStop Color="Red"`
		`                    Offset="1" />`
		`    </LinearGradientBrush>`
		`  </Ellipse.Fill>`
		`</Ellipse>`

RadialGradientBrush	Paints a gradient between a focal point and a circle on the outside of the shape	`<Ellipse>`
		`  <Ellipse.Fill>`
		`    <RadialGradientBrush>`
		`      <GradientStop Color="Blue"`
		`                    Offset="0" />`
		`      <GradientStop Color="Red"`
		`                    Offset="1" />`
		`    </RadialGradientBrush>`
		`  </Ellipse.Fill>`
		`</Ellipse>`

ImageBrush	Paints an image	`<Ellipse>`
		`  <Ellipse.Fill>`
		`    <ImageBrush ImageSource="/foo.jpg" />`
		`  </Ellipse.Fill>`
		`</Ellipse>`

VideoBrush	Paints a MediaElement	`<Ellipse>`
		`  <Ellipse.Fill>`
		`    <ImageBrush SourceName="theVideo" />`
		`  </Ellipse.Fill>`
		`</Ellipse>`

Each property of a XAML element that accepts a brush object can take any of the different types of brushes.

Colors

XAML contains a set of built-in colors. You can use these 141 named colors to specify individual colors, like so:

```
<Grid>
  <Rectangle Fill="Blue"
                  Stroke="Pink" />
</Grid>
```

In most cases, though, named colors end up being insufficient to handle the basics of colors. Since millions of colors are available, XAML needs a way to more effectively specify a color. XAML supports the HTML convention of an RGB hexadecimal string, like so:

```
<Grid>
  <Rectangle Fill="#0000FF"
                  Stroke="#FF0000" />
</Grid>
```

In this format, the pound symbol (#) is followed by a set of hexadecimal numbers that represent the amount of red, green, and blue being used. Both the six- and three-digit formats are supported (e.g., #FF0000 is equivalent to #F00). In addition, XAML extends the HTML syntax to include an eight-character version. In the eight-character version, the first two characters represent a hexadecimal number that indicates the alpha channel (or level of opaqueness):

```
<Grid>
  <Rectangle Fill="#800000FF"
                  Stroke="#C0FF0000" />
</Grid>
```

In this example, the Fill is roughly 50% transparent and the Stroke is approximately 75% opaque (or 25% transparent).

Text

For basic drawing of text, the TextBlock class is the right tool. The Text-Block is a simple container for drawing text. It supports properties for basic font choices such as size, family, weight, foreground color, alignment, and so on:

```
<Grid>
  <TextBlock Text="Hello World"
                  Foreground="White"
                  FontFamily="Segoe WP"
                  FontSize="18"
                  FontWeight="Bold"
                  FontStyle="Italic"
```

```
                    TextWrapping="Wrap"
                    TextAlignment="Center" />
    </Grid>
```

Along with simple text, the `TextBlock` class also supports simple inline formatting using the `LineBreak` and `Run` constructs:

```
<Grid>
  <TextBlock>
    Hello World. <LineBreak />This
    is the second line. The breaking of
    the lines in the XAML are
    <Run Foreground="Red">not significant</Run>.
  </TextBlock>
</Grid>
```

A `LineBreak` indicates where line breaks are going to occur without regard to the `TextWrapping` property. A `Run` is used to wrap some piece of text that needs to be formatted differently than other parts of the `Text-Block`. The `Run` supports the basic properties that a `TextBlock` allows but only applies them to the text inside the `Run` element as shown above. The `TextBlock` is not a control to handle any sort of rich text or HTML-level text handling but will suffice in most cases for text manipulation.

Images

Although the simple vector drawing stack is invaluable to the design of your Windows Phone 7 application, you will always need to use images in your application. The simple `Image` element is used to display images in your application:

```
<Image Source="http://wildermuth.com/images/headshot.jpg" />
```

The `Image` element supports JPEG and PNG files. It does not support GIF files. By specifying the `Source` attribute, the `Image` element shows the picture you specify in the URI of the source. Specifying an Internet URI, the `Image` element will attempt to download the image from the Internet location. The `Source` attribute supports a relative URI as well:

```
<Image Source="headshot.jpg" />
```

By using a relative URI, the `Image` element attempts to download the image from the application itself. You can add an existing image to the Silverlight project by simply selecting Add | Existing Item from the Project menu. Once you have the image as part of the project, it will be packaged with your application. Therefore, you can simply use the relative URI to specify the `Source` attribute. The relative URI is relative to the root of the project. So if you were to put an image in a project folder, the URI would navigate to the path:

```
<Image Source="Images/headshot.jpg" />
```

Storing your images as part of the application is typical for static images (e.g., button icons, backgrounds, etc.).

By default, the `Image` element is set to stretch the image to fit the size of the element. You can stretch images by specifying the `Stretch` attribute. The valid types of stretch include

- **None:** No stretching is performed.
- **Uniform:** Stretches the image, preserving the original aspect ratio, to fit within the frame of the `Image` element. This is the default.
- **UniformToFill:** Stretches the image, preserving the original aspect ratio, to fill the `Image` element. If the aspect ratio of the `Image` element is different from that image, the image will be clipped to accommodate the difference.
- **Fill:** Stretches the image to fill the `Image` element without preserving the aspect ratio.

Figure 3.2 illustrates examples of the different stretch types.

When creating `Image` elements you can simply specify the `Stretch` attribute, like so:

```
<Image Source="Images/headshot.jpg" Stretch="UniformToFill" />
```

Because the `Image` element is just part of the design grammar, you can specify size either by using height and width or by using container properties like any other element (e.g., `Grid.Row/Column`, `Margin`, `Vertical-Alignment`, etc.).

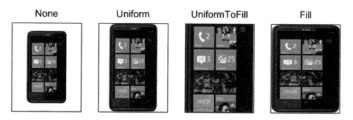

FIGURE 3.2 Image stretching

Transformations and Animations

Now that you have the basic building blocks of designing the look of an application, let's talk about creating the "feel" of an application. The feel of an application is the way it interacts with the user. The level of interaction depends on the nature of the application, but many applications should feel alive to the user. Often this is accomplished with subtle feedback to the user, including changing the look of the UI in reaction to the user's actions or using techniques such as haptic (e.g., vibration) feedback. This feedback is important to help the user know he is doing something. A common example of this is the venerable button object. In a typical desktop operating system, when you move your mouse over a button it changes its look to indicate you're over the button. When you click on it, it changes its look to give you the impression that it is actually pressed (like a real-world button). This feedback ensures that you can feel confident that pressing the button is doing what you expect. Some websites lack this feedback, which simply confuses users (often they don't know what is missing). This is where transformations and animations can help you polish your user interface design.

Transformations

Let's start with transformations. The idea of a transformation is to simply change the way an element is drawn on the screen. Let's take a simple rectangle:

```
<Grid>
  <Rectangle Width="100"
             Height="100"
             Fill="Red" />
</Grid>
```

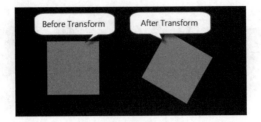

FIGURE 3.3 Transformations in action

As you would expect, this rectangle will be drawn as a simple square. Let's see what happens when we add a transform (by assigning it to the Rectangle's RenderTransform property):

```
<Grid>
  <Rectangle Width="100"
             Height="100"
             Fill="Red">
    <Rectangle.RenderTransform>
      <RotateTransform Angle="30" />
    </Rectangle.RenderTransform>
  </Rectangle>
</Grid>
```

By using a RotateTransform, you ensure that the object can remain a Rectangle, but when drawn, the transformation is applied (as seen in Figure 3.3).

Using a transformation on a single element rarely is the right thing to do; usually a transformation is applied to an entire container to change the look of the container:

```
<Canvas>
  <Canvas.RenderTransform>
    <RotateTransform Angle="30"
                     CenterX="150"
                     CenterY="150" />
  </Canvas.RenderTransform>
  <Ellipse Width="300"
           Height="300"
           Stroke="Black"
           Fill="Yellow"
           StrokeThickness="2" />
  <Ellipse Fill="Black"
           Width="50"
           Height="50"
           Canvas.Left="50"
           Canvas.Top="75" />
```

```
    <Ellipse Fill="Black"
           Width="50"
           Height="50"
           Canvas.Left="200"
           Canvas.Top="75" />
  <Path Stroke="Black"
       StrokeThickness="5"
       Data="M 50,200 S 150,275 250,200" />
</Canvas>
```

In this case the entire smiley face design is rotated (as seen in Figure 3.4).

Table 3.4 describes and provides examples of the different types of transformations.

FIGURE 3.4 Entire container transformed

TABLE 3.4 Transformation Types

Type	Description	Example
RotateTransform	Performs a two-dimensional rotation on an object or object tree	`<RotateTransform Angle="30" />`
SkewTransform	Performs a two-dimensional skew on an object or object tree	`<SkewTransform AngleX="30" AngleY="75" />`
ScaleTransform	Scales an object or object tree	`<ScaleTransform ScaleX="1.5" ScaleY=".75" />`
TranslateTransform	Moves an object or object tree in two dimensions	`<TranslateTransform X="1.5" Y=".75" />`
CompositeTransform	Performs a mix of rotation, skewing, scaling, and translation in a preferred order	`<CompositeTransform Rotation="30" ScaleX="1.5" TranslateX="150" />`

You can use the four basic types of transforms singularly or, if you need to mix transforms (e.g., scale and rotate), you can use `Composite-Transform` to combine multiple transforms.

Animations

Although the idea of animations in applications may make you think of creating the next blockbuster animated feature, that's not what animations are for at all. Animations are simply a way to change properties of XAML elements over time. For example, a simple animation to change the width of a rectangle would look like this:

```
<DoubleAnimation Storyboard.TargetName="theRectangle"
                 Storyboard.TargetProperty="Width"
                 From="50"
                 To="250"
                 Duration="00:00:05" />
```

Animation elements tell a specific property how to change over time. This example shows how to change the width of an element named `theRectangle` from 50 to 250 over five seconds. The attached properties (`Storyboard.TargetName` and `Storyboard.TargetProperty`) hint at the fact that animations are not executed on their own but are housed in a container called a `Storyboard`. For example:

```
<Grid.Resources>
  <Storyboard x:Name="theStory">
    <DoubleAnimation Storyboard.TargetName="theRectangle"
                     Storyboard.TargetProperty="Width"
                     From="50"
                     To="250"
                     Duration="00:00:05" />
  </Storyboard>
</Grid.Resources>
```

The `Storyboard` is embedded in a `Resources` section (usually at the main container or `UserControl` level) and named so that it can be executed and controlled via code. The unit of work for animations is the `Storyboard`. `Storyboard`s can contain one or more animations, but all animations are executed concurrently (not consecutively). Therefore, if we expand this `Storyboard` to include two animations:

```
<Grid.Resources>
  <Storyboard x:Name="theStory">
    <DoubleAnimation Storyboard.TargetName="theRectangle"
                     Storyboard.TargetProperty="Width"
                     From="50"
                     To="250"
                     Duration="00:00:05" />
    <DoubleAnimation Storyboard.TargetName="theEllipse"
                     Storyboard.TargetProperty="Opacity"
                     From="1"
                     To="0"
                     Duration="00:00:03" />
  </Storyboard>
</Grid.Resources>
```

when this storyboard is executed both animations will execute at the same time (again concurrently), even though the animations themselves are against entirely different properties on different objects.

The animations you've seen so far have been of the type Double-Animation. These animations are used because the animation is changing a number (a double value). There are also animations to change colors (ColorAnimation) and vectors (PointAnimation). All three of these animation types change values over a consistent time frame (and are called **timeline animations**). There are also animations called **keyframe animations.** These animations also change properties over time, but the calculation is based on a value at a specific time in the animation. For example:

```
<DoubleAnimationUsingKeyFrames Storyboard.TargetName="theRectangle"
                               Storyboard.TargetProperty="Height">
  <LinearDoubleKeyFrame KeyTime="00:00:00" Value="50" />
  <LinearDoubleKeyFrame KeyTime="00:00:01" Value="150" />
  <LinearDoubleKeyFrame KeyTime="00:00:03" Value="200" />
</DoubleAnimationUsingKeyFrames>
```

There are keyframe animations for each timeline animation (e.g., double, point, and color) but they are named with the UsingKeyFrames post-fix, as shown above. The Storyboard attached properties are still used to signify the target of the animation, but instead of a simple To and From to specify the values of the animation, one or more keyframes are used. For example, the LinearDoubleKeyFrame element specifies at what time the value should be a specific numeric property.

In this case, the height should start at 50 at the start of the animation; move quickly over the first second to 150; and finally slow down and move to 200 over the last two seconds. The interpolation of the values between the keyframes depends on the type of keyframe. In this example the interpolation is linear. You can also use spline and discrete to achieve curved interpolation and stepped interpolation, respectively.

With these tools in hand, you should be able to create the subtle interactive effects that give the user the impression that he is interacting with real-world objects.

XAML Styling

When writing code, it's customary to take common pieces of code and reuse them in a number of ways, including creating base classes, creating static classes, or even creating reusable libraries. XAML has the same need for creating reuse in the design. This reusability, though, is more about creating a consistent look for the application without having to copy the same code over and over. Consider this common XAML:

```
<TextBox x:Name="nameBox"
         FontSize="36"
         FontFamily="Segoe WP"
         FontWeight="Black"
         BorderBrush="Blue"
         Foreground="White"
         HorizontalAlignment="Stretch" />
<TextBox x:Name="emailBox"
         FontSize="36"
         FontFamily="Segoe WP"
         FontWeight="Black"
         BorderBrush="Blue"
         Foreground="White"
         HorizontalAlignment="Stretch" />
```

In this XAML many of the properties are copied from one of the TextBoxes to the other. If we change any of the properties of one, we will have to copy the change to the other to provide consistency of the UI. In addition, the Foreground and BorderBrush are using colors that could or should be part of an overall look and feel. It's likely that we would want the brushes used there to be consistent not only from TextBox to TextBox but also across the entire application. That's where styling and resources come in.

Understanding Resources

The first level of consistency has to do with sharing common resources. When creating an application, you often will want to use common colors or brushes across the application. Silverlight allows you to create objects to be used in more than one area by specifying them in a `Resources` section and identifying the object with an `x:Key` attribute. For example, you could define a `SolidColorBrush` in a `Resources` section like so:

```
<Grid x:Name="LayoutRoot">
  <Grid.Resources>
    <SolidColorBrush x:Key="mainBrush"
                     Color="Blue" />
  </Grid.Resources>
  ...
</Grid>
```

Every class that derives from `FrameworkElement` (which means most XAML elements) supports a collection of `Resources`. We can refer to these named elements using the `StaticResource` markup extension, like so:

```
<Grid x:Name="LayoutRoot">
  <Grid.Resources>
    <SolidColorBrush x:Key="mainBrush"
                     Color="Blue" />
  </Grid.Resources>
  <TextBlock Foreground="{StaticResource mainBrush}"
             Text="Hello World" />
  ...
</Grid>
```

The `StaticResource` markup extension tells the XAML parser to replace the foreground with the main brush. You can use the resource in several places, which isolates it from changes so that later, when you change the main brush to a `LinearGradientBrush`, it cascades down to wherever the `StaticResource` was used.

The `StaticResource` markup extension looks up through the XAML document to find the resource with the correct name (through the hierarchy), and will continue beyond the beginning of the XAML document. Above the XAML document is the App.xaml file in the phone application project. Normally this is where you would place any application-wide resources. For example, if the App.xaml file looked like this:

```
<Application x:Class="PhoneControls.App"
             xmlns="..."
             xmlns:x="..."
             xmlns:phone="..."
             xmlns:shell="...">

  <!--Application Resources-->
  <Application.Resources>
    <SolidColorBrush x:Key="mainBrush"
                     Color="Blue" />
  </Application.Resources>

 . . .

</Application>
```

the `mainBrush` would then be defined at the application level so that any XAML document that wanted to use the brush could do so, like so:

```
<Grid x:Name="LayoutRoot">
  <TextBlock Foreground="{StaticResource mainBrush}"
             Text="Hello World" />
  . . .
</Grid>
```

Although this example shows a brush (which is a very commonly shared resource), it is not limited to only brushes. Any creatable object can be used in this way. In addition, when you want to share these resources across projects you can accomplish this with `ResourceDictionary` objects. **Resource dictionaries** are XAML files that contain shared resources that can be imported into App.xaml using merged dictionaries. These dictionaries can be flat XAML files or can be contained in separate assemblies that are referenced to your phone application. For more information on merged dictionaries, see the documentation.

Understanding Styles

While sharing resources can help you to define a common look and feel, the styling stack extends this idea by allowing you to specify the default properties for controls in a common place. The `Style` object allows you to create these default properties:

```
<Style TargetType="TextBox"
       x:Key="mainTextBox">
```

```
    <Setter Property="FontSize"
          Value="18" />
    <Setter Property="FontFamily"
          Value="Segoe WP Bold" />
</Style>
```

The `Style` object takes the type of object it can be applied to and then a set of `Setter` objects that define default values for properties. In this example, the `FontSize` and `FontFamily` for a `TextBox` are supplied. To apply this style to an object, you can map it to the `Style` property on an element via the `StaticResource` markup extension, like so:

```
<TextBox Style="{StaticResource mainTextBox}" />
```

By setting this `TextBox`'s `Style` property using the `StaticResource` markup extension, the default property values of the `TextBox` will be set using the `Style`. `Styles` are just named resources, so you would typically place them in the App.xaml file along with other resources. In addition, you can use resources inside your styles, like so:

```
<Application.Resources>
  <SolidColorBrush x:Key="mainBrush"
                   Color="Blue" />
  <Style TargetType="TextBox"
         x:Key="mainTextBox">
    <Setter Property="FontSize"
          Value="18" />
    <Setter Property="FontFamily"
          Value="Segoe WP Bold" />
    <Setter Property="Foreground"
          Value="{StaticResource mainBrush}" />
  </Style>
</Application.Resources>
```

In this way, the shared resources can cascade down into the styling stack. In addition, the `Styles` themselves can be cascaded by using the `BasedOn` property:

```
<Application.Resources>
  <SolidColorBrush x:Key="mainBrush"
                   Color="Blue" />
  <Style TargetType="TextBox"
         x:Name="baseTextBox">
    <Setter Property="FontSize"
          Value="18" />
```

```
    <Setter Property="FontFamily"
           Value="Segoe WP Bold" />
  </Style>
  <Style TargetType="TextBox"
        x:Key="mainTextBox"
        BasedOn="{StaticResource baseTextBox}">
    <Setter Property="Foreground"
           Value="{StaticResource mainBrush}" />
  </Style>
</Application.Resources>
```

Finally, styles can be polymorphic. In other words, the `TargetType` may apply to a base class and be applied to all objects of that type. For example:

```
<Application.Resources>
  <SolidColorBrush x:Key="mainBrush"
                   Color="Blue" />
  <Style TargetType="Control"
        x:Name="baseControl">
    <Setter Property="BorderBrush"
           Value="Black" />
  </Style>
  <Style TargetType="TextBox"
        x:Key="mainTextBox"
        BasedOn="{StaticResource baseControl}">
    <Setter Property="FontSize"
           Value="18" />
    <Setter Property="FontFamily"
           Value="Segoe WP Bold" />
    <Setter Property="Foreground"
           Value="{StaticResource mainBrush}" />
  </Style>
</Application.Resources>
```

Since the `TargetType` of the base style was `Control`, it could be used as the `BasedOn` for any controls (or even the `Style` for any control that derived from the `Control` class).

Implicit Styles

While you will often use named styles (and the `StaticResource` markup extension) to tie a `Style` to a XAML element (as shown in the preceding section), you can also create styles that apply to elements by default. These are called **implicit styles.** To create an implicit style, your style must not include a key. For example:

```
<Application.Resources>
  <Style TargetType="TextBox">
    <Setter Property="FontSize"
            Value="18" />
    <Setter Property="FontFamily"
            Value="Segoe WP Bold" />
  </Style>
</Application.Resources>
```

By eliminating the x:Key on the Style, the style will apply to every element of the TargetType (e.g., TextBox). If an element is specifically styled with an explicit (e.g., named) style, the implicit style is completely replaced. Therefore, you can have either an implicit or an explicit style applied to a specific XAML element, not both. In most cases you will have an implicit style for the main style of a control, then specific explicit styles to handle specific use cases for controls.

One big difference in implicit styles is that the TargetType is not polymorphic, so it only applies to the specific type, not derived types. For example, if you create an implicit style of type "Control", it will only apply to XAML elements of the Control class specifically; TextBox and Button elements (which derive from Control) will not use the style at all. The other rules for styles (e.g., using base resources, cascading styles with BasedOn) all apply.

Where Are We?

This chapter introduced you to the basics of the XAML ecosystem in Silverlight for Windows Phone 7. With these concepts down you are ready to start designing the user interfaces for your applications. I have only touched the surface of the nature of XAML, so you should not assume that I described every element and every attribute in this chapter. Use the documentation to fill out your knowledge of XAML in Silverlight.

Whereas this chapter focused on understanding the textual representation of the XAML, the next chapter will introduce you to using controls in your applications to interact with users.

◤ 4 ◼
Controls

While the drawing grammar is useful for designing an exciting and dynamic application for the phone, most of the functionality users expect has to do with interacting with your application. That is where controls come in. **Controls** support direct interaction with users. The type of interaction depends on the control. For example, buttons and sliders use the touch interface; the `TextBox` uses a keyboard (on-screen or hardware). Using controls in your phone applications requires that you think differently about how you build applications. If you simply take any experience you have on the Web or in desktop applications and try to apply it to the phone, your application will not be easy to use. Taking the smaller screen and touch interface into account will help you build compelling applications using controls.

Controls in Silverlight

Controls are no different from any other XAML elements you have seen so far. For example, here is the `TextBox` control:

```
<Grid>
  <TextBox Text="Hello World"
           Height="75"/>
</Grid>
```

FIGURE **4.1** `TextBox` **control example**

This `TextBox` will show up like the drawing elements but will support the user interacting with the control through touch (as evidenced by the cursor shown in Figure 4.1).

Out of the box, Silverlight for Windows Phone SDK 7.1 supports these controls:

- `Button`
- `CheckBox`
- `HyperlinkButton`
- `ListBox`
- `PasswordBox`
- `ProgressBar`
- `RadioButton`
- `RichTextBox`
- `Slider`
- `WebBrowser`
- `TextBox`

These controls represent the main form of interaction with users. While this list is somewhat abbreviated, these controls are specialized to support the Windows Phone touch interface. Most of these controls are built larger than you might imagine (and with large margins) to accommodate users touching them.

Most of the controls in Silverlight fit into one of three categories, which should help you to understand how the controls are expected to work:

- Simple controls
- Content controls
- List controls

Silverlight Controls

If you are coming to this book with existing Silverlight knowledge you may be surprised by the abbreviated nature of the list of controls to be supported. While many of the controls in Silverlight 4 (and the Silverlight Toolkit for Silverlight 4) will work with Windows Phone, Microsoft has not redesigned these controls to be easy to use with the phone. If you need these other controls, they are not forbidden; it is just up to you to change the way they look to conform to the Metro design language as well as make them work sufficiently with the touch-based input that is available on the phone.

The controls Microsoft chose have specific integration with the phone's touch interface. When you start to look at other controls (e.g., `ToolTip`, `Calendar`, etc.), you will see that finding the right functionality for these controls in a touch environment is not simple. Therefore, you may want to stick with the built-in controls until you have a good feel for the way touch affects how users interact with the controls.

Simple Controls

The simple controls include the `TextBox`, `PasswordBox`, `Slider`, and `ProgressBar`. These controls have a simple API in that they do a specific job and look a certain way. They are, in a word, simple:

```
<StackPanel>
  <TextBox Text="Hello" />
  <PasswordBox PasswordChar="*" />
  <Slider Value="5" />
  <ProgressBar IsIndeterminate="True" />
</StackPanel>
```

Using Keyboards

The `TextBox` and `PasswordBox` controls support text input by the user. For devices with physical keyboards this is simple, but for the majority of devices (that don't have physical keyboards) you must use a **software**

FIGURE 4.2 Software input panel (SIP)

input panel (SIP). A SIP is shown when either of these controls has received focus. You can see the default SIP in Figure 4.2 (both the portrait and landscape versions).

The SIP attempts to put the most common keys directly on the keyboard but also supports ways of getting at the rest of the characters. Figure 4.3 shows you these special keys. These include the Shift key (#1 in the figure), the &123 key (#2), and special long-hold keys (such as the period key, #3).

The long-hold keys offer a way to pop up commonly used keys without making the keyboard looked cramped. For example, in Figure 4.4 you can

FIGURE 4.3 Special SIP keys

Figure 4.4 Long-hold keys

see the standard SIP's period key when the user holds it for more than two seconds.

This default look of the SIP is only one of many different layouts that are supported on the phone. When you are building applications that require text input you will want to tell the controls which SIP layout to use. Deciding on the features of the different SIP layouts is important. The faster users can enter data, the happier they will be.

Changing the look of the SIP is as simple as using something called an **input scope.** For example, you can specify that you want to have some chat features (such as a button for smiley faces) by using the chat input scope, like so:

```
<TextBox InputScope="Chat" />
```

This changes the SIP to be friendlier for a simple chat, as shown in Figure 4.5.

The chat input scope adds an emoticon button, as well as an Autocorrect panel, to help users more quickly type what they want to say. The items in the long-hold keys are also customized for the type of task that the input scope specifies. For example, when the user is typing on the SIP using the chat input scope, the exclamation point character is located in the period's long-hold list. But if the user is typing a URL, the colon and slash characters are in the long-hold keys.

While there are a large number of input scopes you can use, for most applications the InputScope values listed in Table 4.1 should help you pick the right one for your use.

FIGURE 4.5 Chat input scope

TABLE 4.1 Common InputScope Values

Input Scope	Layout	Use Case
Default	QWERTY	When entering nondictionary words such as usernames
Text	QWERTY	When entering text that may be helped by autocorrect and capitalization (includes visual indicator of misspelled words)
Chat	QWERTY	When constructing chat messages (where abbreviations are more important) such as Twitter or SMS messages
URL	QWERTY	When entering an Internet URL
EmailSmtpAddress	QWERTY	When entering an email address
TelephoneNumber	12-key	When entering phone numbers

TABLE **4.1** Common `InputScope` Values (*continued*)

Input Scope	Layout	Use Case
Search	QWERTY	When the user wants to enter search phrases (includes visual indicator of mis-spelled words)
`NameOrPhoneNumber`	QWERTY	When entering names (e.g., SMS mes-sages) but quick access to a 12-key layout for phone number entry is desired
Date	QWERTY	When entering dates; both numeric and character dates (e.g., 12/12/2011 and December 12, 2011)
Maps	QWERTY	When entering addresses; simplifies entry by defaulting to numeric entry

`RichTextBox` *Control*

The Windows Phone SDK 7.1 includes a specialized control for displaying formatted text. This control is called the `RichTextBox`. With the `Rich-TextBox` control you can format text by using a simplified markup including paragraph, bold, italic, and hyperlink tags.

The format is meant to provide some level of formatting like HTML text allows without requiring the complexity (or power) of the full HTML stack. For example:

```
<RichTextBox>
  <Paragraph>You can use inline tags to format
    <Bold>bold</Bold> text and even add
    <Italic>italics</Italic>.
  </Paragraph>
  <Paragraph>Using this Markup you can add
    <LineBreak /> line breaks and even add
    <Hyperlink NavigateUri="/SomePage.xaml">hyperlinks</Hyperlink>!
  </Paragraph>
  <Paragraph>Also arbitrary XAML:
    <InlineUIContainer>
      <StackPanel>
        <Ellipse Fill="Red"
                 Width="25"
                 Height="25" />
```

```
        <TextBox />
      </StackPanel>
    </InlineUIContainer>
  </Paragraph>
</RichTextBox>
```

The `RichTextBox` control supports a number of tags, as shown in Table 4.2.

Silverlight Developers

For those of you who are using Silverlight for the desktop, the `Rich-TextBox` on the phone just supports read-only mode.

TABLE 4.2 `RichTextBox` Markup Tags

Tag	Description
Paragraph	Represents the main container for text information. The `RichTextBox` typically contains a collection of `Paragraph` tags.
Bold	Formats the text within the tag to be bold.
Italic	Formats the text within the tag to be italic.
Underline	Formats the text within the tag to be underlined.
Run	Contains unformatted text to add formatting to.
Span	Contains any elements to add formatting to. Cannot contain `Paragraph`, `InlineUIContainer`, or `Hyperlink` tags in current version of Windows Phone.
LineBreak	Inserts an explicit break in the text formatting.
Hyperlink	Used to create arbitrary text that, when clicked, will open a new page.
InlineUIContainer	Allows arbitrary XAML to be inserted inline into the text of the `RichTextBox`.

Content Controls

Content controls specifically allow you to contain arbitrary XAML inside them. The most common of these is the `Button` control. For example, to simply show a text message in a button you could just set the `Content` property, like so:

```
<StackPanel>
  <Button Content="Click Me" />
</StackPanel>
```

You can see that the content of the control ("Click Me") is now inside the button, as shown in Figure 4.6.

But the content can take arbitrary XAML content as well:

```
<StackPanel>
  <Button>
    <Button.Content>
      <StackPanel>
        <Image Source="headshot.png"
               Width="100"/>
        <TextBlock>Hello</TextBlock>
      </StackPanel>
    </Button.Content>
  </Button>
</StackPanel>
```

This results in a button containing XAML, as shown in Figure 4.7.

Notice that the content is inside the button, not replacing the XAML that makes up the button. Setting the `Content` property allows you to

FIGURE 4.6 Simple button with simple content

FIGURE 4.7 Button with XAML content

specify what is inside the button (or in most content controls, what is inside some part of the control). A content control is any control that derives from the ContentControl class. These include Button, CheckBox, Radio-Button, and HyperlinkButton.

List Controls

List controls support showing any arbitrary list of items. They do this by using a property called ItemsSource. This property takes any collection that supports IEnumerable or IList. This means any type of collection (from simple arrays to complex generic collections) is supported by the list controls. The ListBox defined in XAML is pretty standard:

```
<StackPanel>
  <ListBox x:Name="theList" />
</StackPanel>
```

The real trick is when you set some collection to the ItemsSource property:

```
public partial class MainPage : PhoneApplicationPage
{
  // Constructor
  public MainPage()
  {
    InitializeComponent();

    theList.ItemsSource = new string[] { "One", "Two", "Three" };
  }
}
```

Setting the ItemsSource will show the collection and allow individual items to be selected, as shown in Figure 4.8.

Other list controls will follow this same interface (of setting the collection to an ItemsSource) to set the collection. By using these simple control sets, you should be able to create great experiences for your users.

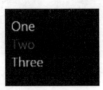

FIGURE 4.8 List box

Phone-Specific Controls

So far all the controls you've seen have existed in earlier versions of Silverlight. But some controls are specifically for use on the phone. The two most obvious are the `Pivot` and `Panorama` controls that allow you to create multipanel controls in a cohesive way. First we'll discuss the `Panorama` control.

Panorama **Control**

The `Panorama` control creates a virtual canvas of several panels that the user can scroll into view as she wants. The `Panorama` control allows you to build these virtual canvases out of one or more panels. You can see an example of a panorama application in Figure 4.9.

The `Panorama` control requires that you add a reference to the `Microsoft.Phone.Controls` assembly. Likewise, you must import the new namespace into your XAML document, as shown here:[1]

```
<phone:PhoneApplicationPage
  x:Class="MyFirstPanorama.MainPage"
  xmlns:ctrls="clr-namespace:Microsoft.Phone.Controls;
          assembly=Microsoft.Phone.Controls"
  xmlns="http://schemas.microsoft.com/winfx/2006/xaml/presentation"
```

FIGURE 4.9 Panorama application

---

1. Note that the namespace has a line break in it before the "assembly" part of the namespace. This is for illustration. In your XAML, the entire contents of the namespace should have no line breaks.

Once you have the new namespace you can use the `Panorama` and `PanoramaItem` elements to create the panorama:

```
<Grid x:Name="LayoutRoot"
        Background="Transparent">
    <ctrls:Panorama Title="my panorama">
      <ctrls:PanoramaItem Header="first">
        <Grid>
          <ListBox />
        </Grid>
      </ctrls:PanoramaItem>
      <ctrls:PanoramaItem Header="second">
        <Grid>
          <ListBox Width="500" />
        </Grid>
      </ctrls:PanoramaItem>
    </ctrls:Panorama>
  </Grid>
```

The `Panorama` element supports a `Title` attribute, which is used to display text that goes across the entire `Panorama` control. Inside the `Panorama` control you can use one or more `PanoramaItem` elements. Each `PanoramaItem` element represents one second in the panorama. The `PivotItem`'s `Header` property controls what is shown above each `PanoramaItem` section. You can see the panorama in Figure 4.10.

The `Panorama` element's `Title` property is labeled #1 in Figure 4.10. You can see that the title is shown across all the panes, so it is never shown in its entirety. The two panorama items in the design are shown as individual panes. The area labeled #2 shows the first `PanoramaItem` element and the area labeled #3 shows the second `PanoramaItem`. Notice that the next panorama item is hinted at on the right side of the screen.

FIGURE 4.10 Panorama explained

Panoramas commonly have a background image that slides behind the panorama as the user moves from one panorama item to the next. To get that behavior, you can set the `Background` element of the panorama using an `ImageBrush`. For example, you would use the following code to paint the background with an image in the .xap file:

```
<ctrls:Panorama Title="my panorama">
  <ctrls:Panorama.Background>
    <ImageBrush ImageSource="/back.jpg"
                Opacity=".2" />
  </ctrls:Panorama.Background>
...
```

While panorama sections (e.g., `PanoramaItem` elements) are meant to take up most of a single screen on the phone, other sections can be larger than a single screen. You can see that the section labeled #1 in Figure 4.11 is larger than a single screen, while the section labeled #2 is sized for a single page. This is consistent with the design paradigm for the phone.

To use landscape sections, you must make a couple of changes. By default, a panorama section will take up most of a single page. To get the larger panes you must both size the `PanoramaItem`'s contents to be as large as necessary (using the `Width` attribute) as well as set the orientation of the `PanoramaItem` to `Landscape`, as shown with the second `PanoramaItem` in the code that follows:

FIGURE 4.11 Landscape sections

```
<ctrls:Panorama Title="my panorama">
  <ctrls:Panorama.Background>
    <ImageBrush ImageSource="/back.jpg"
                Opacity=".2" />
  </ctrls:Panorama.Background>
  <ctrls:PanoramaItem Header="first">
    <Grid>
      <ListBox />
    </Grid>
  </ctrls:PanoramaItem>
  <ctrls:PanoramaItem Header="second"
                      Orientation="Horizontal">
    <Grid Width="750">
      <ListBox />
    </Grid>
  </ctrls:PanoramaItem>
  <ctrls:PanoramaItem Header="third">
    <Grid>
      <ListBox Width="750" />
    </Grid>
  </ctrls:PanoramaItem>
</ctrls:Panorama>
```

The design guidelines specify that you should have no more than four or five sections in your Panorama controls; fewer than that if you are using landscape sections. The general rule of thumb is to have the entire panorama less than 2,000 pixels wide, though the smaller it is the easier it will be for users to understand the intent.

Pivot **Control**

In addition to the Panorama control, there is another phone-specific control called a Pivot control. The Pivot control is also used to show multiple sections, but the Pivot control can handle a larger number of items than the Panorama control. The chief reason for this is that a pivot section takes the entire width of the page instead of having overlapping sections. For example, the phone's Search page uses a Pivot control, as shown in Figure 4.12.

In the Pivot control, there are labels at the top of the page to show both the currently selected section (#1 in Figure 4.12) and other pages that are not currently selected (#2). The currently selected section is usually white with the other sections gray to indicate a difference between the sections. The Pivot control will show the currently selected section's title always

FIGURE 4.12 `Pivot` **control**

at the top left and scroll the rest of the labels to the right (and off the screen typically). The user switches between sections by swiping left or right or by pressing on the section headers to go to that section automatically. As the section is changed, the content area (#3) is changed to reflect that change.

The `Pivot` control is in the same assembly and namespace as the `Panorama` control (`Microsoft.Phone.Controls`), so adding a reference to the assembly and XML namespace is required as you saw earlier for the `Panorama` control. Building a `Pivot` control in XAML is similar to building a `Panorama` control in that the `Pivot` control can take one or more `PivotItem` elements, like so:

```
<ctrls:Pivot>
  <ctrls:PivotItem Header="first">
    <ListBox />
  </ctrls:PivotItem>
  <ctrls:PivotItem Header="second">
    <ListBox />
  </ctrls:PivotItem>
  <ctrls:PivotItem Header="third">
    <ListBox />
  </ctrls:PivotItem>
```

```
  <ctrls:PivotItem Header="fourth">
    <ListBox />
  </ctrls:PivotItem>
</ctrls:Pivot>
```

Like the `PanoramaItem` element, the `PivotItem` uses the `Header` property to specify the label on top of the `Pivot` control. Note that the `Pivot` control does not support a `Title` attribute as the `Pivot` usually does not have a title above the entire page. The preceding XAML results in the page shown in Figure 4.13.

As the user clicks on the headers or swipes, she can go to the other sections. As with the `Panorama` control, the sections loop, so when the user is on the last section, the first section is to the right of the last page, as shown in Figure 4.14.

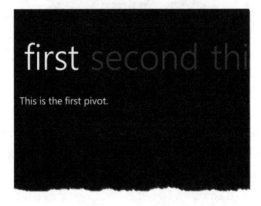

FIGURE 4.13 `Pivot` **control in action**

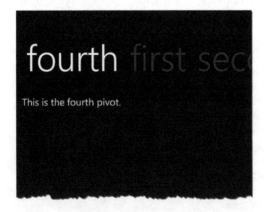

FIGURE 4.14 **Looping pivot sections**

Data Binding

Writing applications for the phone will likely involve data of some sort. Silverlight's support for data binding helps you build your applications in a much more powerful way, but what is data binding exactly? **Data binding** is simply a way to take data from a source (e.g., the value of a property on an object) and show it in a XAML element. If that XAML element is a control, it also supports pushing changes back to the source of the data. While that is a pretty basic explanation, the explanation is correct. Data binding is profoundly simple, and because it is simple, it is powerful.

Simple Data Binding

At the most basic level, a **binding** is a connector to pull data from a data source and put it in an element's property. As you can see in Figure 4.15, a binding is in the middle of the data and the element (`Control`).

Bindings are defined as a markup extension. For example, here is a `TextBox` bound to the `Name` property:

```
<TextBox Text="{Binding Name, Source={StaticResource myData}}" />
```

The `Binding` markup extension first takes a path to the property to be bound to and then a number of optional elements. As you can see in this example, the binding is pulling from a resource object called `myData`. When the data binding happens, it takes the source of the data binding and uses the property path (the `Name` in this case) to navigate to the data it needs to put in the `TextBox`'s `Text`. The path to the property must be a

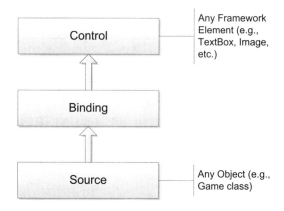

FIGURE 4.15 Simple data binding

public property as it uses reflection to call the getter of the public property to access the data from the source.

Having to specify the source for bindings is relatively rare, though, because when the source of an object changes you would have to change it in a number of places. For example, this XAML would show a simple editor for some data:

```
<StackPanel>
  <TextBlock>Name</TextBlock>
  <TextBox Text="{Binding Name, Source={StaticResource myData}}" />
  <TextBlock>Phone Number</TextBlock>
  <TextBox Text="{Binding Phone, Source={StaticResource myData}}" />
  <TextBlock>BirthDate</TextBlock>
  <TextBox Text="{Binding BirthDay, Source={StaticResource myData}}"
/>
</StackPanel>
```

Imagine that if the source changed, all of the `TextBox`es would need to be rebound. Instead, data binding uses a property called `DataContext`, which simply allows for the source of the data binding to exist along the hierarchy of the XAML. For example, if the `DataContext` were set at the `StackPanel`, all the controls that attempt data binding inside the `StackPanel` would get their data from the `DataContext` instead of needing specific sources:

```
<StackPanel DataContext="{StaticResource myData}">
  <TextBlock>Name</TextBlock>
  <TextBox Text="{Binding Name}" />
  <TextBlock>Phone Number</TextBlock>
  <TextBox Text="{Binding Phone}" />
  <TextBlock>BirthDate</TextBlock>
  <TextBox Text="{Binding BirthDay}" />
</StackPanel>
```

When the bindings pull their data they will look for a source, and when they don't have one they'll search for the first non-null data source in the hierarchy. In this case they will find it at the `StackPanel` level and use that and the source for the data binding. The search for a `DataContext` will continue up the hierarchy until a valid `DataContext` is found. This walking of the XAML tree is not limited to the current XAML document. If the data binding is happening inside a control that is used on another XAML document, it will continue up through all the parents until it exhausts the entire object tree.

Data binding supports three modes, as described in Table 4.3.

TABLE 4.3 Data Binding Modes

Type	Description	Example
OneTime	Pulls data from a source once.	`<TextBox Text="{Binding Name, Mode=OneTime}" />`
OneWay	(Default) Pulls data from a source. As the source's data changes, can pull those changes into the control.	`<TextBox Text="{Binding Name}" />`
TwoWay	Pulls data from a source and pushes changes back to the source as the data changes (normally on the control losing focus).	`<TextBox Text="{Binding Name, Mode=TwoWay}" />`

The pushing and pulling of changes is performed via reflection so that all binding modes work with any .NET class. There are no requirements for that class to work with data binding. The one exception to that is if you want changes to the source object itself to be reflected in the controls via data binding. For that to work, your source classes must support a simple interface called `INotifyPropertyChanged`. When the source data changes, it notifies the binding that the data has changes that will cause the binding to reread the data and change the data in the control, as shown in Figure 4.16.

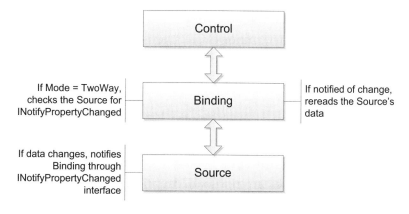

FIGURE 4.16 Changes in the source

Using a DataTemplate

As you saw earlier in this chapter, list controls can show any list that supports IList or IEnumerable, but that's only part of the story. List controls also support the ability to use DataTemplates to customize the look of individual items in the list. You can use a DataTemplate to specify the ItemTemplate to use arbitrary XAML to define what is contained in a ListBox. For example:

```
<ListBox ItemsSource="{StaticResource theData}">
  <ListBox.ItemTemplate>
    <DataTemplate>
      <StackPanel>
        <TextBlock Text="{Binding Name}" />
        <Image Source="{Binding ImageUrl}" />
      </StackPanel>
    </DataTemplate>
  </ListBox.ItemTemplate>
</ListBox>
```

As this ListBox creates its individual items, it will use the Data-Template as a factory to create the XAML that is contained inside. The DataContext for the created XAML becomes the individual item to be shown in the ListBox so that the data binding inside the DataTemplate just works.

Improving Scrolling Performance

When binding against collections on the phone, you must be aware of the implications of how large amounts of data can affect the performance of the scrolling of list controls (e.g., ListBox, ScrollViewer, Items-Source, etc.). There are two minor tweaks you can make to improve this performance: image creation and scroll handling.

For image creation, you can decide how images are actually loaded and decoded. Under the covers of the Image's source is a constructed object called a BitmapImage. The BitmapImage has a property called CreationOptions in which you can specify when an image is loaded. By default, the CreationOptions specify that images should be **delay-loaded.** This means images are not loaded until they can be seen on the surface of a page. In addition to delay-loading the image, you can also specify that an image is loaded on the background thread. Specifying these two things can improve the overall performance of images in a collection. To specify this you would need to break out the `Image.Source` property and set a `BitmapImage` in the control template, like this:

```
<ListBox ItemsSource="{StaticResource theData}">
  <ListBox.ItemTemplate>
    <DataTemplate>
      <StackPanel>
        <TextBlock Text="{Binding Name}" />
        <Image>
          <Image.Source>
            <BitmapImage UriSource="{Binding ImageUrl}"
              CreateOptions="DelayCreation,BackgroundCreation" />
          </Image.Source>
        </Image>
      </StackPanel>
    </DataTemplate>
  </ListBox.ItemTemplate>
</ListBox>
```

You can see in this example that the `Image` is using the verbose XAML syntax to set the `Source` to a `BitmapImage` object. The `UriSource` takes the same binding the earlier example used for the `Source`. Finally, the `CreateOptions` is set to both `DelayCreation` and `BackgroundCreation` to improve the performance of loading this image in the user interface.

Additionally, you can improve the performance of scrolling by allowing the object responsible for scrolling in lists (the `ScrollViewer` class) to decide whether scrolling should be handled by the operating system (the default) or by the control. Depending on the size of the scrolling region, giving scrolling responsibility to the control can improve the user's experience. To change this, you can specify the `ScrollViewer.ManipulationMode` attached property on any list control (e.g., `ListBox`, `ItemsControl`, `ScrollViewer`, etc.) to the value of `Control`, as shown here:

```
<ListBox ItemsSource="{StaticResource theData}"
        ScrollViewer.ManipulationMode="Control">
<ListBox.ItemTemplate>
  <DataTemplate>
    <StackPanel>
      <TextBlock Text="{Binding Name}" />
      <Image>
        <Image.Source>
          <BitmapImage UriSource="{Binding ImageUrl}"
            CreateOptions="DelayCreation,BackgroundCreation" />
```

```
        </Image.Source>
      </Image>
    </StackPanel>
  </DataTemplate>
</ListBox.ItemTemplate>
```

Binding Formatting

During the data binding process, you have the opportunity to format the data directly in the binding. Several binding properties allow you to specify what happens during binding. These include StringFormat, Fallback-Value, and TargetNullValue. Each can impact what the user sees during data binding. For example, you can include these as additional properties inside the binding, as shown here:

```
<TextBox Text="{Binding ReleaseDate,
                  StringFormat=d,
                  FallbackValue='n/a',
                  TargetNullValue='n/a'}" />
```

The StringFormat property is used to specify a .NET format string to be used during binding. This can be any .NET format string that matches the type. If the .NET format string contains spaces, you should surround it with single quotes. The StringFormat is used both to format the data when pushing it to the control as well as to parse the data going back to the source.

The FallbackValue is used to show a value when the binding fails. A binding can fail if it does not find a source (e.g., source or data context is null) or when the source does not have a valid property to bind to.

Finally, the TargetNullValue is used to indicate that the binding succeeded, but the value of the bound result is null.

Element Binding

You can also create bindings that allow you to create a link between two XAML elements. This is called **element binding.** To use element binding, you can specify the ElementName as part of the binding syntax, like so:

```
<Slider Minimum="10"
        Maximum="36"
        x:Name="fontSizeSlider" />
<TextBox FontSize="{Binding Value, ElementName=fontSizeSlider}"
         Text="Make It Grow" />
```

The size of the font in the `TextBox` is being set based on the value of the `Slider` (named `fontSizeSlider`). In this way, you can use elements in the XAML to supply data to other elements in the XAML. A more conventional use of element binding is to set the data context of a container based on the selected value of a control, like so:

```
<ListBox ItemsSource="{Binding Games}"
         x:Name="theList" />
<StackPanel DataContext="{Binding SelectedItem, ElementName=theList}">
  <TextBlock>Name</TextBlock>
  <TextBox Text="{Binding Name}" />
  <TextBlock>Phone Number</TextBlock>
  <TextBox Text="{Binding Phone}" />
</StackPanel>
```

In this example, the `StackPanel` is setting its `DataContext` to whatever item is selected in the `ListBox`. In this way, you can show and/or edit the data in the `StackPanel` based on the selection of the `ListBox`.

Converters

Data binding takes properties from objects and moves them into properties on controls. At times, the types of the properties will not match or will need some level of manipulation to work. That is where converters come in. **Converters** are stateless classes that can perform specific conversions during the binding process. In order to be a converter, a class must implement the `IValueConverter` interface. This interface has two methods, `Convert` and `ConvertBack`, to allow for conversions in both directions during binding. For example, a simple converter to make dates show up as short date strings looks like so:

```
public class DateConverter : IValueConverter
{

  public object Convert(object value,
                        Type targetType,
                        object parameter,
                        CultureInfo culture)
  {
    if (targetType == typeof(string) &&
        value.GetType() == typeof(DateTime))
    {
      return ((DateTime)value).ToShortDateString();
    }
```

```
    // No Conversion
    return value;


}

public object ConvertBack(object value,
                         Type targetType,
                         object parameter,
                         CultureInfo culture)
{
  if (targetType == typeof(DateTime) &&
    value.GetType() == typeof(string))
  {

    DateTime newDate;

    if (DateTime.TryParse((string)value, out newDate))
    {
      return newDate;
    }
  }

  // No Conversion
  return value;
  }
}
```

Converters are created as resources (usually at the application level) like so:

```
<Application x:Class="PhoneControls.App"
             xmlns="..."
             xmlns:x="..."
             xmlns:phone="..."
             xmlns:shell="..."
             xmlns:my="clr-namespace:PhoneControls">
  <Application.Resources>
    <my:DateConverter x:Key="dateConverter" />
  </Application.Resources>
  ...
</Application>
```

By creating the converter at the application level, you can use it throughout the application. Finally, we can now use the converter directly in our data binding, like so:

```
<StackPanel DataContext="{Binding SelectedItem, ElementName=theList}">
  <TextBlock>Name</TextBlock>
  <TextBox Text="{Binding Name, Mode=TwoWay}" />
  <TextBlock>Phone Number</TextBlock>
  <TextBox Text="{Binding PhoneNumber, Mode=TwoWay}" />
```

```
    <TextBlock>Phone Number</TextBlock>
    <TextBox Text="{Binding ReleaseDate,
                    Mode=TwoWay,
                    Converter={StaticResource dateConverter}}"
/>
</StackPanel>
```

During the conversion of the underlying data (in this case a `Date-Time`), the `DateConverter` class is used. When moving the data from the source to the control, `Convert` is called; when the data is pushed back to the source, the `ConvertBack` method is called. As in this example, converters are often used just for formatting and not real conversion.

Data Binding Errors

By design, data binding errors do not cause exceptions. This behavior is desirable because the source of a data binding may enter a valid and invalid state a lot during the life of your application. Let's take the example we saw earlier where we have a list of controls bound to the `SelectedItem` of a `ListBox`. When there is no selection in the `ListBox` the data binding is failing. Throwing an exception in that case would be the wrong thing to do. So, as a developer, you will need a way to actually see data binding failures. Luckily you can see them pretty clearly in the Visual Studio Output window. When running your application, you can use the View menu to show the Output window, as shown in Figure 4.17.

When data binding fails, it adds a debug message to the Output window. For example, if you used the wrong path in a `Binding` (e.g., `Title` instead of `Name`) you could see this in the Output window, as shown in Figure 4.18.

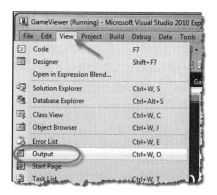

FIGURE 4.17 Output window

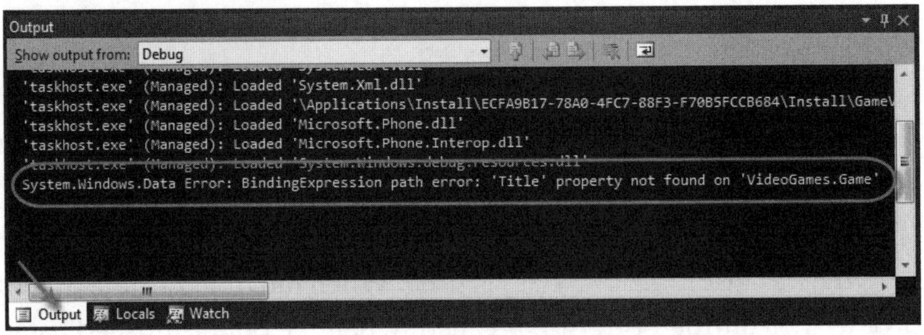

FIGURE 4.18 Binding error shown in the Output window

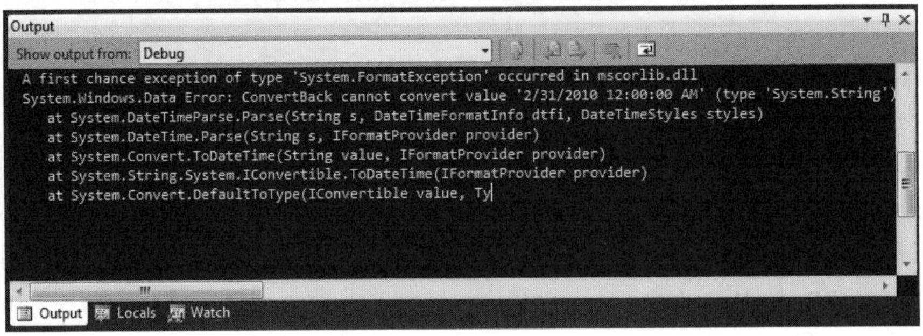

FIGURE 4.19 Conversion error shown in the Output window

Bad paths aren't the only binding errors that show up in the Output window; bad conversions (really, any exceptions) do as well. For example, if a user attempted to enter a bad date (e.g., 2/31/2010) into a date field, the Output window would show that error, too, during data binding (as seen in Figure 4.19).

Control Templates

Although property-based styling is very powerful, it may not let you change the look of the controls in dramatic ways. That is where control templates come in. Every control in Silverlight has XAML that defines how a control is drawn. For example, the humble button uses this XAML to draw itself:

```
<Grid Background="Transparent">
    ...
    <Border x:Name="ButtonBackground"
            BorderBrush="{TemplateBinding BorderBrush}"
            BorderThickness="{TemplateBinding BorderThickness}"
```

```
          Background="{TemplateBinding Background}"
          CornerRadius="0"
          Margin="{StaticResource PhoneTouchTargetOverhang}">
    <ContentControl x:Name="ContentContainer"
                  ContentTemplate="{TemplateBinding
                                              ContentTemplate}"
                  Content="{TemplateBinding Content}"
                  Foreground="{TemplateBinding Foreground}"
                  HorizontalAlignment="{TemplateBinding
                                     HorizontalContentAlignment}"
                  Padding="{TemplateBinding Padding}"
                  VerticalAlignment="{TemplateBinding
                                     VerticalContentAlignment}"
/>
  </Border>
</Grid>
```

Even though you think of controls as atomic elements, there is XAML inside the control to define the look and feel of the control. Control templates are used to redefine this XAML for any control.

Control templates are part of the style that is applied to a control. The `ControlTemplate` is the value of the `Template` property of any control. For example:

```
<Style x:Key="ButtonStyle1"
       TargetType="Button">
  <Setter Property="Template">
    <Setter.Value>
      <ControlTemplate TargetType="Button">
        <Grid Background="Transparent">
          ...
        </Grid>
      </ControlTemplate>
    </Setter.Value>
  </Setter>
</Style>
```

The contents of the `ControlTemplate` contain the XAML that the particular control should use (i.e., `Button` in this case). As you define the XAML that makes up the look of a particular control, you can use a markup extension called a **template binding** to pull the value of a property into your XAML. For example:

```
<ControlTemplate TargetType="Button">
  <Grid Background="Transparent">
```

```
<Border x:Name="ButtonBackground"
        BorderBrush="{TemplateBinding BorderBrush}"
        Background="{TemplateBinding Background}">
    ...
</Border>
  </Grid>
</ControlTemplate>
```

By using template bindings, the natural value of the property will be used as the value inside the control template. This value could come from the default value in the control or from a style setter, or it could specifically be set on an instance of the control. Template bindings allow you to use whatever property value is supposed to be shown in the control.

For some controls, the XAML must have certain elements to make sure the control still works. For example, the `WebBrowser` control requires that part of the XAML be a `Border` element named `PresentationContainer`. That way, the control knows where to show the Web content. This contract between you and the control author is called a **template part** (many controls do not have any template parts). The template parts that are required are documented as attributes on the controls themselves (as shown in Figure 4.20).

The structure of the XAML in a control template represents the look of the control, but in addition you can specify the feel of an application. The feel of an application is the way the control can interact with the users. For

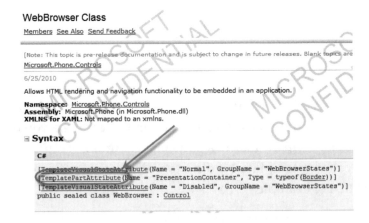

FIGURE 4.20 `TemplatePart` attribute

example, the `Button` class changes its appearance when pressed to give feedback to the user that she has correctly pressed the button. This is how the feel of an application works.

You can create the feel of an application using a structure called the **visual state manager.** With the visual state manager you can define animations that represent a state the control can be in. The control uses the visual state manager to go into specific states as the user interacts with it. As the application designer, you can define these states to change the way the control interacts with the user. These states are broken up into groups so that a single control can be in more than one state. For example, two of the visual state manager groups that the `TextBox` has are `CommonStates` (which represent states such as `Disabled` and `ReadOnly`) and `FocusStates` (which represent whether the control has focus or not). The groups define a set of states where only one state can be active at a time. For example, you can have your control be `Disabled` and `Focused`, but not `Disabled` and `ReadOnly`. The states in a group are mutually exclusive.

To define the groups and states for the visual state manager, the XAML can contain a `VisualStateManager.VisualStateGroups` property (as an attached property):

```
<ControlTemplate TargetType="Button">
  <Grid Background="Transparent">
    <VisualStateManager.VisualStateGroups>
      <VisualStateGroup x:Name="CommonStates">
        <VisualState x:Name="Normal" />
        <VisualState x:Name="MouseOver" />
        <VisualState x:Name="Pressed">
          <Storyboard>
            <ObjectAnimationUsingKeyFrames
                Storyboard.TargetProperty="Foreground"
                Storyboard.TargetName="ContentContainer">
              <DiscreteObjectKeyFrame KeyTime="0"
                                      Value="{StaticResource
                                              PhoneBackgroundBrush}" />
            </ObjectAnimationUsingKeyFrames>
            ...
          </Storyboard>
        </VisualState>
        <VisualState x:Name="Disabled">
          <Storyboard>
            <ObjectAnimationUsingKeyFrames
```

```
                    Storyboard.TargetProperty="Foreground"
                    Storyboard.TargetName="ContentContainer">
             <DiscreteObjectKeyFrame KeyTime="0"
                                     Value="{StaticResource
                                             PhoneDisabledBrush}" />
           </ObjectAnimationUsingKeyFrames>
           ...
         </Storyboard>
       </VisualState>
     </VisualStateGroup>
   </VisualStateManager.VisualStateGroups>
   ...
 </Grid>
</ControlTemplate>
```

As this example shows, the `VisualStateManager.VisualState-Groups` attached property contains one (or more) `VisualStateGroup` objects. Inside the group is a list of `VisualState` objects that represent a storyboard that shows how to go to a specific state. The empty `Visual-State` objects mean the state should look exactly like the natural state of the object.

The visual states and groups that a control supports are also specified as attributes on the control classes, as shown in Figure 4.21.

When creating your own control templates you will need to be aware of the template parts and template visual states, as that is the contract between you and the control author. You must implement these states and parts if you expect the controls to continue to operate correctly.

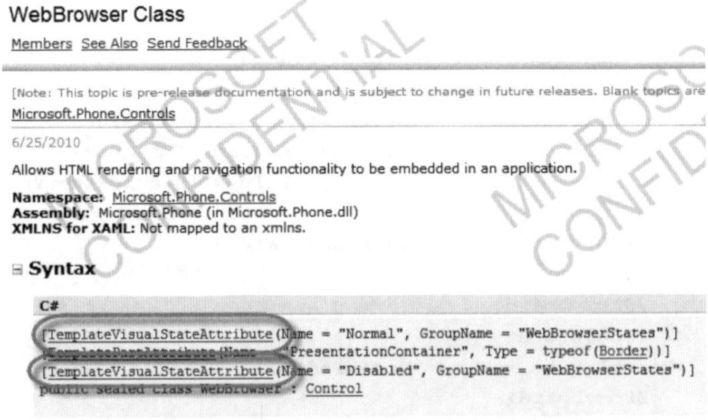

FIGURE 4.21 `TemplateVisualState` attribute

Silverlight for Windows Phone Toolkit

In addition to the controls that are part of the Windows Phone SDK 7.1, there is another download called the Silverlight Toolkit. The version of the toolkit for the phone includes a set of controls specific to the phone to help you create compelling applications. You should install the Silverlight Toolkit to add these controls (and other features you'll learn about in subsequent chapters) to your applications. You can download the Silverlight for Windows Phone Toolkit directly from the CodePlex site at http://silverlight.codeplex.com. You can just download the installer or you can opt to download the entire source code if you're interested in how these controls have been built.

The Silverlight for Windows Phone Toolkit includes the following controls:

- `AutoCompleteBox`
- `ContextMenu`
- `DatePicker`
- `TimePicker`
- `ListPicker`
- `LongListSelector`
- `PerformanceProgressBar`
- `ToggleSwitch`
- `ExpanderView`
- `PhoneTextBox`
- `WrapPanel`

We will discuss these controls in the subsections that follow.

`AutoCompleteBox` Control

First up is the `AutoCompleteBox`. The purpose of this control is to allow you to suggest selections as a user types into the control. The control is styled to look just like the `TextBox`, but as the user types, the control can show a list of possible options. For example, in Figure 4.22 when the user types "S" the control shows a pop up with all options that start with the letter *S*.

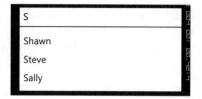

FIGURE 4.22 `AutoCompleteBox` **example**

The control is a list control, so you can just assign an arbitrary list to the `ItemsSource` property:

```
theBox.ItemsSource = new string[]
{
  "Shawn",
  "Steve",
  "Sally",
  "Bob",
  "Kevin"
};
```

There are a number of ways to customize the control using XAML attributes, including specifying whether text completion is enabled and what type of filtering to support (`StartsWith` is the default, but you can have the suggestions based on `Contains` or `Equals` as well):

```
<toolkit:AutoCompleteBox x:Name="theBox"
                    IsTextCompletionEnabled="True"
                    FilterMode="Contains" />
```

A very common approach with the `AutoCompleteBox` is to support a list of values that are not known at development time. This is how the Bing and Google search boxes work on the Web. You can achieve this by handling the `TextChanged` event and then filling in the `ItemsSource` as the text changes:

```
// Constructor
public MainPage()
{
  InitializeComponent();

  theBox.TextChanged += new RoutedEventHandler(theBox_TextChanged);
}
```

```
void theBox_TextChanged(object sender, RoutedEventArgs e)
{
  // Go retrieve a list of items from a service
}
```

ContextMenu **Control**

The purpose of the ContextMenu control is to allow users to long-click on parts of your application to get a list of options. The control by default shows itself large enough to be very obvious to the user. You can see the ContextMenu control in action with three menu items and a separator in Figure 4.23.

The structure of the ContextMenu control consists of a ContextMenu element with a collection of one or more items inside the ContextMenu. There is only a single level of menu items, so no submenus are supported. The two types of items are MenuItem elements and Separator elements:

```
<toolkit:ContextMenu>
  <toolkit:MenuItem Header="Add" />
  <toolkit:MenuItem Header="Remove" />
  <toolkit:Separator />
  <toolkit:MenuItem Header="Cancel" />
</toolkit:ContextMenu>
```

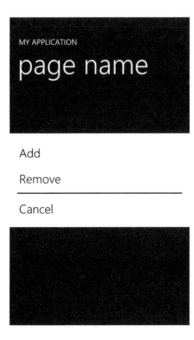

FIGURE 4.23 ContextMenu **example**

To add a context menu to a XAML element, you will use the `Context-Menu` attached property to apply it to your design, like so:

```
<Grid>
  <toolkit:ContextMenuService.ContextMenu>
    <toolkit:ContextMenu>
      <toolkit:MenuItem Header="Add" />
      <toolkit:MenuItem Header="Remove" />
      <toolkit:Separator />
      <toolkit:MenuItem Header="Cancel" />
    </toolkit:ContextMenu>
  </toolkit:ContextMenuService.ContextMenu>
  ...
</Grid>
```

Once the menu is attached to the element, a user touch-hold will cause the menu to be displayed. The individual `MenuItem` elements can launch code either via an event or via a `Command` binding:[2]

```
...
<toolkit:MenuItem Header="Add"
                  Click="MenuItem_Click" />
<toolkit:MenuItem Header="Remove"
                  Command="{Binding RemoveCommand}" />
...
```

DatePicker **and** TimePicker **Controls**

If you've designed desktop or Web applications before, you probably are used to finding a calendar control to allow users to pick dates. The problem with a calendar control on the phone is that the interface is not very touch-friendly. Instead, the phone supports a control for picking dates: `DatePicker`. Using the `DatePicker` is as simple as using the XAML element:

```
...
<TextBlock>Pick Date</TextBlock>
<toolkit:DatePicker />
...
```

---

2. Command binding is not covered in this book but is a useful technique for separating the XAML from the code. Please see the Silverlight documentation for the `ICommand` interface for more information.

FIGURE 4.24 Date picking user interface

The `DatePicker` looks like a simple `TextBox` that accepts a date. The difference is that when a user taps the control it launches a full-screen date picking user interface, as shown in Figure 4.24.

The date picking user interface allows the user to pan and flick to pick the date. This interface works really well with a touch interface. You will notice that this interface shows an Application bar at the bottom, but the icons are missing. In order to use this control, you need to add the icons to your project manually. The images are retrieved by way of a relative path to the icons, so your .xap file must contain the icons. You first need to create a directory in your application, called Toolkit.Content. Once you have that directory, you need to add the icons from the toolkit directory in the Icons folder. The Icons folder is stored in the toolkit directory:

```
%PROGFILES%\Microsoft SDKs\Windows Phone\v7.1\Toolkit\{Version}\Bin\Icons
- e.g.
%PROGFILES%\Microsoft SDKs\Windows Phone\v7.1\Toolkit\Aug11\Bin\Icons
```

After adding these icons, you should ensure that these files are added as "Content," as shown in Figure 4.25.

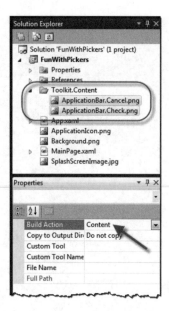

FIGURE 4.25 Setting icons as "Content"

Once these files are part of the project, the icons show up as expected. The `DatePicker` control defaults to `DateTime.Now`, which is why the current date is shown. You can specify a date using the `Value` property:

```
<toolkit:DatePicker Value="04/24/1969" />
```

The `TimePicker` works in exactly the same way as the `DatePicker` but it uses the time portion of the `DateTime` structure:

```
...
<TextBlock>Pick Time</TextBlock>
<toolkit:TimePicker Value="12:34 PM" />
...
```

The user interface for picking the time is similar to the `DatePicker`, but allows you to specify the time instead, as shown in Figure 4.26.

`ListPicker` **Control**

I know developers love `ListBoxes`. For the phone, sometimes the `List-Box` is just the wrong tool. For very short lists of options, `ListBoxes` take up too much screen real estate. As an alternative, the toolkit gives you the `ListPicker`.

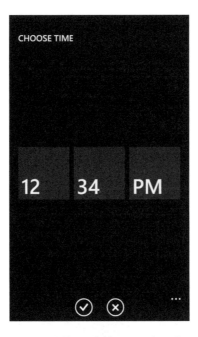

FIGURE 4.26 Time picking user interface

The `ListPicker` control is a good solution when you have a short list from which the user must select one item. In fact, the `ListPicker` could also be used to replace radio buttons. The `ListPicker` shows the currently selected item in a box much like a `TextBox`, as shown in Figure 4.27.

When the user touches the `ListPicker`, it opens in one of two ways. If the list is short (five items or less), it expands the control to show the options, as shown in Figure 4.28.

If the list has more than five options, it pops up a full-screen list of options to choose from, as shown in Figure 4.29.

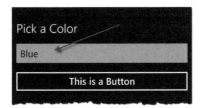

FIGURE 4.27 `ListPicker` example (closed)

Figure 4.28 `ListPicker` example (opened)

Figure 4.29 `ListPicker` example (full screen)

To create a `ListPicker`, you can just create it as a simple XAML element, like so:

```
<TextBlock Style="{StaticResource PhoneTextLargeStyle}"
Text="Pick a Color" / >
<toolkit:ListPicker x:Name="thePicker" />
<Button Content="This is a Button" />
```

Since the `ListPicker` is a list control, you can use the `ItemsSource` to specify the list:

```
...
thePicker.ItemsSource = new string[]
{
  "Blue",
  "Green",
  "Red",
  "Orange",
  "Purple",
  "Cyan",
  "Brown",
  "Gray",
  "Light Green"
};
...
```

`LongListSelector` Control

As an alternative to using the `ListBox` for very long lists, the toolkit also supplies you with a control that lets users look at large numbers of options. The phone uses this when you pick a phone number from your address book. Because the list of people could be quite long, it categorizes the people by the first letter of their first or last names. While it supports three different modes, the real gem is the ability to have a pop-up list of groups to help users locate the items they're looking for. Although this type of control is used for the address book, that application uses the first letter for grouping, but it's completely up to you how you decide to group the objects in the `LongListSelector`. For example, Figure 4.30 shows a list of games grouped by genre.

This shows the groups to the user and lets her tap on the group to pop up an overlay of groups to help her navigate large lists effectively, as shown in Figure 4.31.

In order to use this control, you have to lean on data binding to set up three elements of your control.

- **ItemTemplate:** This is the contents of the individual data for an item in the `LongListSelector` (i.e., a "Game" in the preceding example).

FIGURE **4.30** `LongListSelector` **with groups**

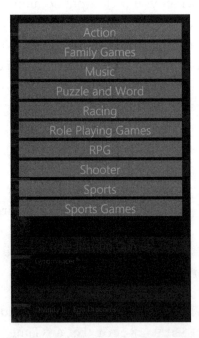

FIGURE **4.31** `LongListSelector`**'s pop-up groups**

- **GroupHeaderTemplate:** This is the item above the list of grouped objects (i.e., the "Genre" in the preceding example) shown in the main UI of the `LongListSelector`.
- **GroupItemTemplate:** This is the display for a group in the pop-up.

It is common to use the same template for the group item and the group header. Here is an example of the XAML to create a `LongListSelector`:

```xml
<Grid x:Name="LayoutRoot"
      Background="Transparent">

  <Grid.Resources>

    <DataTemplate x:Key="letterTemplate">
      <Border Background="{StaticResource PhoneAccentBrush}"
              Margin="4">
        <TextBlock Text="{Binding GroupName}"
                 VerticalAlignment="Center"
                 HorizontalAlignment="Center"
                 Style="{StaticResource PhoneTextGroupHeaderStyle}" />
      </Border>
    </DataTemplate>

  </Grid.Resources>

  <toolkit:LongListSelector x:Name="theSelector"
              GroupHeaderTemplate="{StaticResource letterTemplate}"
              GroupItemTemplate="{StaticResource letterTemplate}">
    <toolkit:LongListSelector.ItemTemplate>

      <DataTemplate>
        <StackPanel Orientation="Horizontal"
                    Background="Transparent">
          <Image Height="75"
                 Source="{Binding ImageUrl}" />
          <TextBlock Text="{Binding Name}"
                   Style="{StaticResource PhoneTextNormalStyle}"/>
        </StackPanel>
      </DataTemplate>

    </toolkit:LongListSelector.ItemTemplate>
  </toolkit:LongListSelector>
</Grid>
```

Notice first that the XAML stores a `DataTemplate` for the group in a resource. It does this so that we can use the same template for both the `GroupHeaderTemplate` and the `GroupItemTemplate`. Next, the `ItemTemplate` is specified inline. This works much like the `ListBox` examples earlier in this chapter. The grouping is what really changes the nature of how this control works.

When you apply data to the control, you can just assign the `Items-Source` like any other list control. The problem is that to make the control work it expects your data to be in a very specific format. This format is a collection of groups. A group is just a collection that often has something that describes it (such as the name of the genre in our example). Silverlight already has something like this called the `IGrouping<T,T>` interface. You might try to just use the grouping semantics in LINQ to accomplish this, like so:

```
// THIS DOES NOT WORK

// Use LINQ to Group
var games = new GameList();
var qry = from g in games
          orderby g.Genre, g.Name
          group g by g.Genre into genres
          select genres;

// Bind the collection of Groups into the control
var result = qry.ToList();
theSelector.ItemsSource = result;
```

This LINQ query sorts the games by the name of the genre and the name of the game, then groups that result into collections of names by genre. This sounds very much like what we need for the control. Unfortunately, the underlying class that handles the grouping does not support data binding because the name of the group is not public.[3] To solve this you can use a simple wrapper for the grouping, like so:

```
public class Group<T> : List<T>
{
  public Group(IGrouping<string, T> group)
  {
    GroupName = group.Key;
    this.AddRange(group);
  }
```

---

3. As noted in the Data Binding section, data binding uses reflection and Silverlight does not support non-public reflection.

```
    public string GroupName { get; set; }
}
```

This class just adds a public property for the name of the group and constructs itself from an `IGrouping<T,T>` object that LINQ uses (though it assumes a string-based key). Remember, not only does this group have a `GroupName` property to identify the group to the user, but it also *is* a collection of the underlying objects. This is the data format this control requires, which is why we needed this class. This way, you can modify the LINQ query to construct these instead of returning the raw `IGrouping<T,T>` interface:

```
// Use LINQ to Group
var qry = from g in games
          orderby g.Genre, g.Name
          group g by g.Genre into genres
          select new Group<Game>(genres);
```

This works because our grouping data template uses the `GroupName` we specified in the `Group<T>` class to do the data binding:

```
<DataTemplate x:Key="letterTemplate">
  <Border Background="{StaticResource PhoneAccentBrush}"
          Margin="4">
    <TextBlock Text="{Binding GroupName}"
               VerticalAlignment="Center"
               HorizontalAlignment="Center"
               Style="{StaticResource PhoneTextGroupHeaderStyle}" />
  </Border>
</DataTemplate>
```

The `LongListSelector` supports other templates and properties to control the way you present this to the user, but understanding the basics of how to get a simple version of the control working will help you get started with the control.

PerformanceProgressBar[4] Control

The built-in `ProgressBar` control has some known performance problems, including the fact that it renders on the wrong thread (making it

_____

4. This control continues to be available in the Silverlight for Windows Phone Toolkit but these same changes were made to the built-in control, therefore making this control unnecessary except for backward compatibility with Windows Phone 7.0 applications.

appear jumpy) and the fact that the animations continue even if the control is stopped. This means if you want to use a progress bar in your application, you should use the `PerformanceProgressBar` control from the toolkit instead. This control exists because the toolkit has much shorter release cycles and pushing a new progress bar to all phones would require an operating system update. Releasing this in the toolkit means users can get the performance gains without waiting for the next phone update.

You can simply use this control as your progress bar, instead of the built-in one, wherever you might expect to put the `ProgressBar`, like so:

```
<ProgressBar IsIndeterminate="{BindingIsBusy}" />
```

Replace this with the toolkit version, like so:

```
<toolkit:PerformanceProgressBar IsIndeterminate="{BindingIsBusy}" />
```

`ToggleSwitch` Control

Although the phone includes a `CheckBox` control, clicking on a box to enable something is not necessarily a good touch-based metaphor. In its place is a toolkit control called a `ToggleSwitch`, as shown in Figure 4.32.

The `ToggleSwitch` allows users to either tap the switch to change its value or actually slide it to the right to enable the option or to the left to disable it. The `ToggleSwitch` is made up of three sections, as shown in Figure 4.33.

The `Header` property controls what is in the section labeled #1 in Figure 4.33. The `Content` property controls what is in the section labeled #2. And the switch itself is shown in the section labeled #3. The `Content` of

FIGURE **4.32** `ToggleSwitch` **example**

FIGURE **4.33** `ToggleSwitch` **components**

the control is usually changed as the state of the control is changed (e.g., the default is On for checked and Off for unchecked). You can see the creation of a `ToggleSwitch` in the following XAML:

```
<toolkit:ToggleSwitch Header="This one is enabled"
                      Content="On"
                      IsChecked="true"/>
<toolkit:ToggleSwitch Header="This one is disabled"
                      Content="On" />
```

`ExpanderView` **Control**

The limited size of the display on the phone means that you may want to conserve the space on the screen as much as possible. One control that will help is the `ExpanderView` control. This control allows you to set up content that is hidden except when the user clicks on the header to show the hidden content. The control consists of a header and content that can be shown when the control is tapped (see Figure 4.34).

To use the control, you need an instance of the `ExpanderView` in your XAML. There are two parts to the control: the header and the items. The header is the part of the control that is always shown and the items contain the content that is shown once the user taps on the header. For example:

```
<toolkit:ExpanderView Header="Click here to Expand">
  <TextBlock>This is hidden by default</TextBlock>
</toolkit:ExpanderView>
```

The `Header` property can be text (as is shown above) or it can be a more complex control using the expanded XAML syntax:

```
<toolkit:ExpanderView x:Name="theExpander">
  <toolkit:ExpanderView.Header>
    <TextBlock FontWeight="Bold"
               Margin="4">Click here to expand</TextBlock>
  </toolkit:ExpanderView.Header>
</toolkit:ExpanderView>
```

FIGURE 4.34 `ExpanderView` in action

For the content, you can just include a list of controls to show simple content or you can use the `ItemsSource` to specify a collection (much like the way a `ListBox` works):

```
public partial class MainPage : PhoneApplicationPage
{
  // Constructor
  public MainPage()
  {
    InitializeComponent();

    expander.ItemsSource = new string[]
      { "Blue", "Red", "Green", "Orange" };
  }
}
```

Like the `ItemsSource` property, you can also use data templates (like the `ItemTemplate`) to control what each item looks like in the `Expander-View`. You should think of the `ExpanderView` as similar to other list controls (detailed earlier in this chapter).

PhoneTextBox **Control**

The built-in `TextBox` is very useful but is missing some features that would make it more useful in some scenarios. It would be nice if the `Text-Box` supported hints (e.g., showing a text watermark of what belongs in a text box before it contains text), length indication, pressing the Return key to create multiple lines of text, and icons to perform actions on the text box. The `PhoneTextBox` fills those needs. Figure 4.35 shows the `PhoneText-Box` when it does not have focus. Notice the watermark ("Enter Tweet") and the action icon on the right that you can hook up events to.

To specify these, you can simply instantiate the control and specify the `Hint` attribute and the `ActionIcon` attribute as necessary:

```
<toolkit:PhoneTextBox x:Name="tweetText"
  ActionIcon="Toolkit.Content/ApplicationBar.Cancel.png"
  Hint="Enter Tweet" />
```

FIGURE 4.35 PhoneTextBox **with the** Hint **and** ActionIcon **shown**

The PhoneTextBox includes an event to use when the ActionIcon is tapped. This event is called ActionIconTapped:

```
public partial class MainPage : PhoneApplicationPage
{
  // Constructor
  public MainPage()
  {
    InitializeComponent();

    tweetText.ActionIconTapped += new
      EventHandler(tweetText_ActionIconTapped);
  }

  private void tweetText_ActionIconTapped(object sender, EventArgs e)
  {
    tweetText.Text = "";
  }
}
```

This example shows clearing the PhoneTextBox when the user clicks the cancel icon. This is a common usability improvement over the standard TextBox.

The PhoneTextBox also supports the ability to show the user how many characters she has typed. As the user types in the control, it can show the number of characters (as well as the maximum number of characters that will be allowed), as shown in Figure 4.36.

The length indicator is supported by setting the LengthIndicator-Visible attribute to True. The LengthIndicatorThreshold attribute is also used to indicate how many characters have to be shown before the indicator is shown. This allows you to not show the indicator until the user is approaching the maximum number of characters. Finally, the Displayed-MaxLength attribute is used to indicate the maximum number in the indicator. Note that this does not limit the length of the field, but simply shows the maximum number in the indicator (which is why it's called Displayed-MaxLength). You can see the XAML where these are specified below:

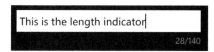

FIGURE 4.36 PhoneTextBox's length indication support

```
<toolkit:PhoneTextBox x:Name="tweetText"
                      DisplayedMaxLength="140"
                      LengthIndicatorVisible="True"
                      LengthIndicatorTheshold="20" />
```

Lastly, you can also indicate that the control should support pressing the Return key to expand the text box to include multiple lines of text. When the user uses this functionality, pressing Enter on the keyboard (or the SIP) will expand the `PhoneTextBox` to include the multiple lines (see Figure 4.37).

You specify this by using the `AcceptsReturn` attribute as shown here:

```
<toolkit:PhoneTextBox x:Name="tweetText"
                      AcceptsReturn="True" />
```

`WrapPanel` **Layout Container**

The last XAML element that is included in the toolkit is a new layout container called the `WrapPanel`. This element is not a control but a layout container (e.g., `Grid`, `StackPanel`). The purpose of the `WrapPanel` is to lay out elements left to right, and when they don't fit horizontally it wraps them onto a new line. For example, if you place nine buttons in a `Stack-Panel` like so:

```
<StackPanel>
  <Button Content="1" />
  <Button Content="2" />
  <Button Content="3" />
  <Button Content="4" />
  <Button Content="5" />
  <Button Content="6" />
  <Button Content="7" />
  <Button Content="8" />
  <Button Content="9" />
</StackPanel>
```

the `StackPanel` will simply stack them vertically, as shown in Figure 4.38.

This is the first line
This is the second
And finally a third!

FIGURE **4.37** PhoneTextBox's AcceptReturn **functionality**

FIGURE **4.38** Buttons in a `StackPanel`

But if you change the XAML to replace the `StackPanel` with a `Wrap-Panel`, like so:

```
<toolkit:WrapPanel>
  <Button Content="1" />
  <Button Content="2" />
  <Button Content="3" />
  <Button Content="4" />
  <Button Content="5" />
  <Button Content="6" />
  <Button Content="7" />
  <Button Content="8" />
  <Button Content="9" />
</toolkit:WrapPanel>
```

the `WrapPanel` will stack the items horizontally and then wrap to a "new line" when the items no longer fit, as shown in Figure 4.39.

You can also change the `Orientation` attribute to `Vertical` to have the control stack vertically:

```
<toolkit:WrapPanel Orientation="Vertical">
  <Button Content="1" />
  <Button Content="2" />
  <Button Content="3" />
```

FIGURE **4.39** Buttons in a `WrapPanel`

```
    <Button Content="4" />
    <Button Content="5" />
    <Button Content="6" />
    <Button Content="7" />
    <Button Content="8" />
    <Button Content="9" />
</toolkit:WrapPanel>
```

This is shown in Figure 4.40.

Where Are We?

The control set for Windows Phone is quite extensive. Combining the included SDK controls with the controls that Microsoft has released as part of the Silverlight for Windows Phone Toolkit will give you a compelling way to build your applications. The most important lesson here is to keep in mind that you're building applications for the phone. Touch is the first-class citizen. This means you have to let go of your old biases of what controls to use where, and try to understand the differences in both touch and real estate. One control that is missing that most developers lean heavily on is a data grid of some sort. The reason for that is that grids and phones don't work well together. Mixing lists, controls, and screen navigation instead of relying on a large canvas of data will help the user more aptly use your application and improve your chances of creating a great and popular application!

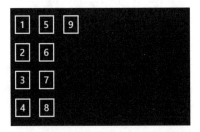

FIGURE 4.40 **Buttons in a vertical** WrapPanel

5

Designing for the Phone

U p to this point you have learned about the phone, worked your way through a simple walkthrough of an application, and learned the basics of XAML. Now you should be starting to think about the nature of the application you want to write. While it is easy to think of the phone as just another screen for Silverlight to exist on, it's not nearly that simple. In this chapter we will talk about the nature of designing for Windows Phone to help you make those hard decisions.

The Third Screen

Microsoft has pushed an idea it calls "three screens and a cloud"[1] since the earliest announcements of Windows Phone (and possibly before). Essentially this is the idea that an application or service should support three fundamental user experiences: computer, TV, and phone. While this idea of three screens that are all supported by a common infrastructure has evolved (e.g., are tablets considered phones or computers?), it still represents a strong story for you to determine what your phone applications should allow the user to do.

Your job in designing the experience for the phone involves more than fitting your Web/desktop experience onto the small screen; it really

---

1. http://tinyurl.com/3screensandthecloud

involves crafting what a user will want to do on the device. For example, let's assume you work for a bank. Should the phone experience include doing things like downloading bank statements? Probably not. But users *will* want to be able to check balances and perhaps see several days' worth of transactions. The Web story for your application may be very feature-driven, whereas the phone version should pick and choose the right experience for the form factor and use cases.

But the experiences you think about are not just a subset of Web experiences; they might be very different. Consider the Foursquare website (http://foursquare.com) shown in Figure 5.1.

The user experience on the website is more heavily geared toward seeing the status of a user; seeing the mayorships, history, and badges is a common task on the website. As a user, I might also be interested in some of the

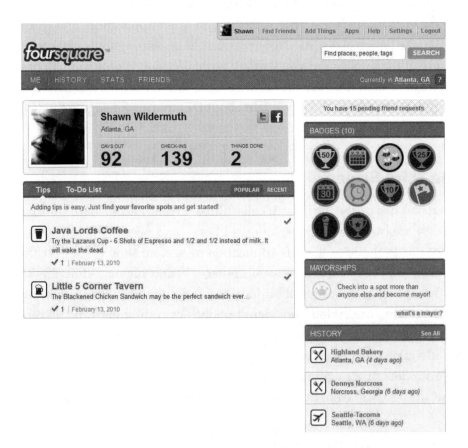

FIGURE 5.1 Foursquare.com

additional data presented to me in this larger format. In fact, there is quite a lot of functionality here that I may be interested in. But just how much of this is interesting on a smaller device? The user's attention span is hampered on a smaller device, so deciding what a user will do (and how long the user is expected to stay in your application) becomes crucial to a successful third screen for your offering. Let's take that site and see how the phone screen might help us pick some functionality (as shown in Figure 5.2).

The problem with this approach is that it's unlikely that picking a phone-sized part of the app would make much sense. Your next thought may be to try to pack all that functionality in with a panorama application (as shown in Figure 5.3).

While Windows Phone would allow this, it's important for you to look at what your third screen will be used for. How long do you expect users

Figure 5.2 Phone-sized app

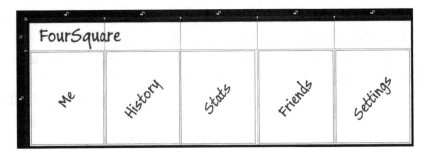

FIGURE 5.3 Panorama application

to work with the application? Even if users want all that functionality, is a large scrollable app the right idea? When I look at Foursquare, I am most interested in check-in (something that is an uncommon task on the website) and seeing where my friends are checked in; the rest of the functionality is purely "nice to have." In fact, the open source Foursquare application for Windows Phone[2] decided on this paradigm instead of being drawn into any of these mistakes in user experience (as shown in Figure 5.4).

FIGURE 5.4 A sample Foursquare on Windows Phone

---

2. http://4square.codeplex.com/

Determining the right experience for your users is your first challenge. Once you have a sense of what you should accomplish, the next challenge is to create an application that works well on a device.

It Is a Phone, Right?

Developing the right strategy for determining what your application does is only half the battle. The other half is to understand that you are working with a phone. Why does it matter that you're writing for a phone? It matters because the hardware, performance characteristics, and usability are completely different from a desktop or website. For example, the Windows Phone Application Certification Requirements[3] for the Windows Phone OS 7.1 has limitations about consuming memory.

> ### ■ Memory Consumption
>
> The application must not exceed 90MB of RAM usage. However, on devices that have more than 256MB of memory, an application can exceed 90MB of RAM usage. The `DeviceExtendedProperties` class can be used to query the amount of memory on the device and modify the application behavior at runtime to take advantage of additional memory. For more information, see the `DeviceExtended-Properties` class in MSDN.

Therefore, you will need to design your application to work well within limitations on a number of fronts:

- Limited screen real estate
- Limited CPU speeds
- Limited battery life
- Limited memory
- Completely different input mechanism (e.g., touch versus mouse)

If you're a seasoned mobile developer, none of this is new to you. But if you're a developer who is coming from the Web or desktop world, you

---

3. http://shawnw.me/n0gbMh

have to change your thinking substantially to fit your applications onto the phone. You will need to start with a clean slate and think about resources in a whole new way.

In addition, on the phone users run apps for very different reasons and expect very different experiences. Typically an application (not necessarily games) is run frequently, but for very short periods of time. And the user experience is very touch-driven instead of keyboard- and mouse-driven. Designing an application to meet this different set of requirements means you really need to dig in to how you expect the application will be used.

Deciding on an Application Paradigm

The Windows Phone style (i.e., Metro) makes some specific recommendations about several styles of applications that look like the rest of the phone. When you design your application you will have to look at these styles and determine if any of them make sense with your application. Alternatively, you can just start from scratch and create a new workflow for your application without regard for Metro. While these application styles are typical, it is not a requirement that your application follow any of these usability paradigms. In fact, some of the built-in apps already do this (e.g., Music and Video).

Much more typical than having a single paradigm for your application is to determine the right mix of these UI metaphors to use in your applications. All Windows Phone applications use a navigation pattern in which you can have multiple pages that support the Back button on the phone. This pattern should match users' expectations as it is borrowed from the Web pattern. A simple app may be made up of several different pages that can be navigated to and from. For example, Figure 5.5 shows a simple blood sugar monitoring app that mixes the different paradigms.

When you look at the design of your application you will need to consider the different ways to create individual pages using the different styles of control, but you should never get away from the idea that Windows Phone revolves around page navigation. Of course, you can build your application as a single page, but that would likely be a very simple application, such as the Moon Phaser app shown in Figure 5.6.

Let's look at some of the page design styles.

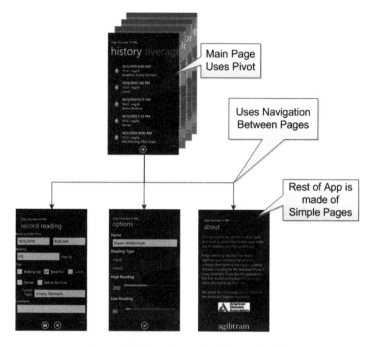

Main Page
Uses Pivot

Uses Navigation
Between Pages

Rest of App is
made of
Simple Pages

FIGURE 5.5 Sample application navigation

FIGURE 5.6 Single-page Windows Phone application

Panorama

The basic construction of the panorama page is an endless canvas of several panes. For example, let's take a simple application with a couple of different parts of the application, as shown in Figure 5.7.

This page will have two pieces of functionality ("first item" and "second item"). The panorama will show the first pane and hint at the second pane. So, this application in the emulator will look like Figure 5.8.

FIGURE 5.7 Sample panorama application

FIGURE 5.8 Panorama in the emulator

In the emulator you can see that the first pane is shown, but it also hints that there is more to be seen. If the user swipes to the left (to move to the right), the second piece of functionality appears. While the user is on the second pane, the first pane is shown again as a preview on the right. This is the magic of the panorama in that it implies an infinite canvas for the different parts of the application. This avoids the typical problem you would have with applications composed of several parts (or panes), and having to develop a way to navigate to them. The background image also is stretched over the panes and slides along with the panes (though at a different speed to give a parallax look to the control). Although this example shows two panes, it is common for a panorama to have more panes. The general rule of thumb is that you should use no more than five panes for a panorama.

Many of the built-in applications use the panorama style to enable different parts of their functionality (e.g., People, Images, Music, and Video). It may be seductive to try to make your application use a panorama since other parts of the phone use that style, but you should only pick this type of application when you really do have multiple pieces of separate functionality. Each pane of the panorama is not meant to be a master-detail type of page. The individual panes are separate pieces of functionality, similar to the Foursquare example in Figure 5.3. Panorama applications also tend to require more horsepower from the phone as all panes are live throughout the lifecycle of the application.

Pivot

Originally designed for Zune (the software and the device), the pivot-style page is a key style that you will see on Windows Phone. It is used in the Music picker (in the Music and Video section) and is a common way to think of displaying the same information in different "views" (i.e., by artist, by album, etc.).

The pivot is similar to the panorama, but use of the pivot is more straightforward. The pivot is also based on the idea of an infinite canvas, but unlike the panorama, the pivot is made of individual tabs and does not require that the next pane shows as a preview. The user can either swipe or click on the headings (the next heading is previewed) to move to that pane. Figure 5.9 shows a typical pivot application.

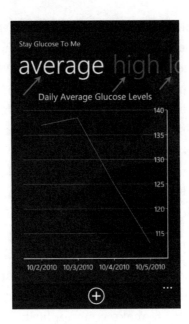

FIGURE 5.9 Pivot example

With the pivot, each pane takes up an entire page. The pivot works in a lot of places where a panorama would be too large and cumbersome. The biggest difference between a panorama and a pivot is that the pivot can handle a larger number of panes. As stated earlier, a panorama application really should be no more than five panes. You can use more than five panes in a pivot, but you cannot use an unlimited number of them; in addition, you should have a very good use case to use a lot of pivot panels.

The pivot style of application is a good choice when you have the same or similar data that needs to be displayed in different views. For example, in the blood monitoring app, where there is a pivot each page shows the data from the first page in different ways (different graphs or just a list of the data), as shown in Figure 5.10.

While the basic metaphor is about showing the same information in different ways on the pivot, you are ultimately in control and could have different functionality on each pivot page. What is important is to not surprise your users.

FIGURE 5.10 Pivot pages

Panorama or Pivot?

In principle you can think of `Pivot` and `Panorama` controls as similar in functionality (e.g., they both show panes of information). In general their use is pretty different. The `Panorama` control is typically used for a small number of panes and encourages a "discovery" type of experience in that the previewing of other panes means users will discover the rest of the functionality more naturally. But this pattern breaks down after four or five panes. The `Panorama` control also is heavier as there is no control over the display of the panes.

The `Pivot` control, on the other hand, is generally for when you want to show the same or similar data in different ways. For example, a common use of the `Pivot` control is in the Music library where you can view the music by artist, genre, and so on. The same information is in different panes but is organized differently.

■ Common Mistake

You should never use the `Panorama` and `Pivot` controls for master-detail interfaces (e.g., one pane used for selection and the rest of the panes used for details of the selected item).

Simple Pages

Sometimes the different page styles just get in the way. That's where typical pages come in. While Microsoft is encouraging some design paradigms with the Metro style, it is not required that you pick one of them. Whether you're building a news application, a casual game, or even a custom video player, you can build it the way you want. The Metro design language should be your guide and should not deter you from creating the application that users will love.

Microsoft Expression Blend

The Windows Phone SDK 7.1 includes a free version of Microsoft's Expression Blend tool for creating your Windows Phone application. You should get comfortable with this tool as it is an important way to design your applications. Blend has been around for a while for use with Silverlight applications, so it has had time to mature before the birth of Windows Phone. For someone who is building apps for Windows Phone, the inclusion of Blend is free (unlike how it is for typical Silverlight development).

Creating a Project

When starting a new application in Blend, you will be presented with a dialog that includes several different project types, as shown in Figure 5.11.

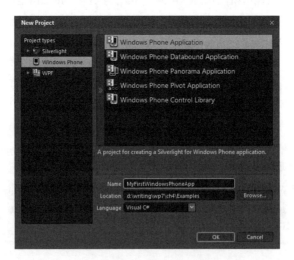

FIGURE 5.11 Blend New Project dialog

TABLE 5.1 New Project Types in Blend

Project Type	Description
Windows Phone Application	Creates a simple, page-based application for Windows Phone
Windows Phone Databound Application	Creates a list-based application with sample date prewired into the user interface for Windows Phone
Windows Phone Panorama Application	Creates a `Panorama` control-based application for Windows Phone
Windows Phone Pivot Application	Creates a `Pivot` control-based application for Windows Phone
Windows Phone Control Library	Creates a library to hold reusable controls that are used in Windows Phone applications

There are separate project types for different starting project types. The first four of these project types are for creating new Windows Phone applications and the last one is for a library to hold controls for other applications to use. Table 5.1 shows the different application projects.

A Tour around Blend

Creating a new application will generate a new project with a base user interface on which you can start your design. This is usually the starting point for designing with Blend. To begin, let's look at the standard user interface layout of Blend to get comfortable with the different elements. The starting user interface contains several elements, as shown in Figure 5.12.

The basic interface of Blend is made up of five main areas (labeled as #1 through #5 in Figure 5.12):

1. **Toolbar:** Commonly used tools live here.
2. **Common panels:** These include Project, Assets, and other panels.
3. **Objects and Timeline panel:** This is the basic layout/navigation panel.
4. **Artboard:** This is the basic design surface for Blend.
5. **Item panels:** These include Properties, Resources, and Data panels.

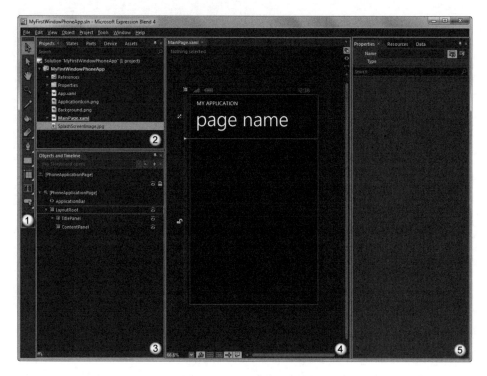

FIGURE **5.12 Blend user interface**

As you start using Blend you'll need to become pretty familiar with how each section can help you. The first area to get comfortable with is the toolbar. The toolbar contains many of the basic tools you will use to create your design. You can see the toolbar (and its submenus) in Figure 5.13.

The basic toolbar is divided into five sections. Some of the individual toolbar buttons support holding them down to show a list of other toolbar buttons (as shown in the items labeled A–F in the figure). You can tell that a toolbar button supports multiple options by the small rectangle in the lower right-hand side of the button. The different toolbar sections are as follows:

1. **Selection tools:** Selection and Direction Selection
2. **View tools:** Hand and Zoom
3. **Brush tools:** Eye Dropper, Paint Bucket, Gradient tool, and Brush Transform (see A)

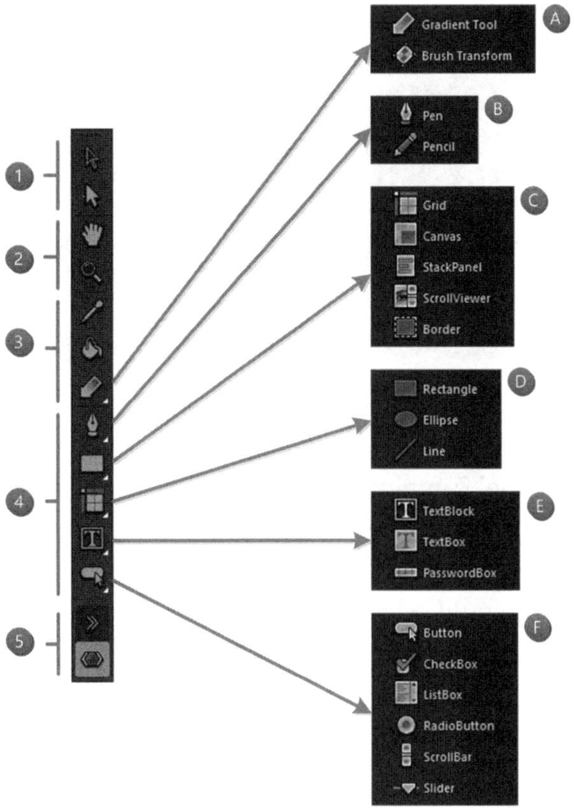

FIGURE 5.13 Blend toolbar

4. **Object tools:** A variety of tools to create different objects on the design surface (as shown in B–F)

5. **Asset tools:** The Asset button and Last Asset Used button

The Common Panel section of the Blend UI contains several commonly used panels. The two that you will interact with the most are the Projects panel and the Assets panel. The other three panels have specific uses that we will discuss later in this chapter.

The Projects panel provides a list of all the projects and files in a Blend project. If you are a designer and are new to Blend, one of the biggest changes you will need to get comfortable with is the idea that you are working with a project with multiple files, not just a single design file. The Projects panel (shown in Figure 5.14) shows all the files in the current project.

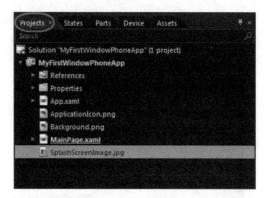

FIGURE 5.14 Projects panel

The Projects panel contains the solution file that contains one or more projects. In this example there is only one project, but you could have additional projects listed, such as libraries of shared designs or code. This panel is also where you would add other assets (e.g., fonts, images, etc.) to the project. Everything in the project can be merged into the resultant phone application.

Another important panel is the Assets panel. Whereas the toolbar buttons give you access to the most common design elements (drawing shapes, containers, and controls), the Assets panel includes all the design elements you can add to a Windows Phone application. The Assets panel is made up of three sections, as shown in Figure 5.15.

As you will see in other parts of Blend, the panel starts with a search bar (#1 in Figure 5.15). Because the Assets panel can contain quite a lot of controls and other assets, searching for assets is often the quickest way to find the asset you are looking for. The categories on the left (#2) separate the different types of assets by individual categories. Notice the small triangles that indicate that some of the asset categories have subcategories. By making a selection in a category, you will show all the controls that fit in that category (or subcategory). As you can see in Figure 5.15, the categories are a mix of asset type (e.g., Media, Controls, and Shapes) as well as location (e.g., Locations and Project). This means an asset could be in more than one category. When a category is selected, the asset list (#3) shows the controls in that particular category or subcategory.

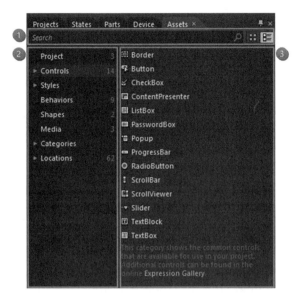

Figure 5.15 Assets panel

The next section of the user interface to look at is the Objects and Timeline panel. This panel lets you review the hierarchy of the objects in your design as well as facilitates the animation design process. Dealing with your object hierarchy is the main purpose of this panel; we will discuss the animation features later in this chapter. For managing the object hierarchy, the Objects and Timeline panel has several key parts, as shown in Figure 5.16.

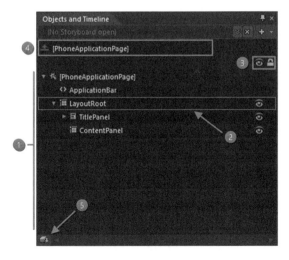

Figure 5.16 Objects and Timeline panel

The object hierarchy (#1 in Figure 5.16) shows all the objects in your design. In this example you can see the `Page`, the `ApplicationBar`, and the `Grid` (called `LayoutRoot`), which contains a couple of other objects. This object hierarchy is the same as the hierarchy of the XAML. The symbol on the left-hand side of each object indicates the object type (they match the icons on the toolbar and Assets panel). For named members, the name is shown in the hierarchy; if an object is not named it will be the object type surrounded by brackets (as seen for the `PhoneApplicationPage`).

When adding new elements to your design, it can be helpful to know what the current container is (which will tell you where new elements will be added). You can see the current container directly in the Objects and Timeline panel because that container will have a blue rectangle around it (#2 in Figure 5.16).

For every element in the object hierarchy, the panel allows you to decide whether to hide or lock the object. The area toward the top right (#3 in Figure 5.16) defines a column for both hiding and locking an object (or a container and its children). Clicking the icons to the right of the objects in the hierarchy will change whether they are shown or locked on the design surface. These changes are purely design-time and have no effect on the runtime look and feel of the objects.

As you move from different objects in your design, the name and icon (#4) give you the ability to move back up to the last editing space. When we talk about styling using Blend, I will show you how this control comes into play.

Lastly, the icon on the lower left of the screen (#5) gives you the ability to flip the object hierarchy order. This button only affects how the hierarchy is displayed in the Objects and Timeline panel. While XAML is written top-down (like HTML) the object hierarchy will show that order. If you click the icon it will flip this order. This reverse order is easier for people who are used to Photoshop or other design tools where the drawing order looks bottom-up.

Moving on to the artboard, shown in Figure 5.17, the main element (#1) of the artboard is much like any other design tool (even Microsoft Paint) in that it allows you to drop and manipulate objects directly. It uses the same handles metaphor as most other design tools as well (e.g., the small

FIGURE 5.17 Artboard

squares on the sides and corners allow you to grab and resize anything on the artboard). Only the selected object(s) on the artboard will show its handles.

On the top of the artboard (#2) are document tabs. For each document (e.g., XAML file) you have open, a tab is created. Clicking on individual tabs allows you to switch to that document. The view buttons (#3) allow you to switch from the design view to a XAML textual view, or to a split view showing both the design and the XAML.

Immediately below the document tags is an indicator of the selected item on the design surface. This may be a drop down (#4) for objects that have contextual menu items (e.g., controls) or it may be a flat box with the type of control selected. This area will also show a breadcrumb of the different subobject design surfaces (e.g., when editing a style or template) to ease the navigation back to the original design surface.

Lastly, the artboard has a zoom drop-down box (#5) to allow you to change the zoom level, and a number of design option buttons that can be toggled on or off to show annotations, show gridlines, and enable rendering of effects.

The last section of the main Blend user interface is the Item Tools section. This section is made up of several important panels, as shown in Figure 5.18.

This section provides tabs for several different panels including (typically) Properties, Resources, and Data (#1). The main panel you will use while you are designing your Windows Phone application is the Properties panel. The main section (#2) of the Properties panel is the list of properties. If you are coming from Visual Studio, the organization and look of the Properties panel is quite different. The properties are sorted into logical groups—for example, Brushes (#3), Appearance, and so on. These groups are collapsible sections that will help you find related properties. Near the end of most of the groups will be a further collapsed section (#4) of less commonly used properties. You can click this arrow icon to open more properties.

FIGURE 5.18 Item Tools panel

FIGURE 5.19 Searching in the Properties panel

Near the top of the Properties panel is an indicator of the name and type of control (#5). If you want/need to name an element, you can simply type it in here. Under the name/type section is the search bar (#6) that allows you to search for property names. This is very useful for finding property names as most of the controls have very large numbers of properties. The search is a straight substring match, so, for example, typing "vis" in the search bar will result in both "Visibility" and "IsHitTestVisible," as shown in Figure 5.19.

Lastly, the Properties panel also has a switch (#7) to change between showing properties and events. Now that you have taken a basic tour of the Blend user interface, let's look at some common tasks using Blend for your design.

Blend Basics

Much like any activity, learning Blend is a matter of performing tasks and repeating them. The more you do, the more the muscle memory of Blend will take over and allow you to create great applications.

Layout

The most basic layout task you will accomplish is simply drawing elements on the screen. With a new application project, if you draw a control (e.g., Button) on the page, you will notice that the size of the control will be shown as you drag, as shown in Figure 5.20.

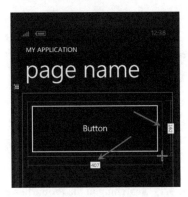

FIGURE 5.20 **Dragging a new control**

Once you let go of the mouse, you will notice that the control will include eight handles for resizing the control. Just outside these handles you'll see lines that radiate outward from the control with small numbers near them. The usual way that Silverlight (and Blend) handle layout is by converting your object size to a margin from the side of the container (in this case a grid). So in Figure 5.21, you can see a "25" near the top and both sides of the new button. This indicates that the top, right, and left margins

FIGURE 5.21 **Margin and alignment layout**

are 25 in size. Near each number is a small closed chain icon that indicates that the object is tied to these edges of the container. The bottom of the button has a dotted line that leads to an open chain icon that indicates that the button is not tied to the bottom. Because it is not tied to the bottom of the container, the margin doesn't appear either (since the margin wouldn't matter in this case). Using this metaphor of chaining to an edge can let you infer the alignment (top vertical aligned and stretch aligned horizontally). You can see the actual alignment and margin in the Properties panel that match this information. You can change these margins and alignments either directly on the artboard or by using the Properties panel.

Most of the dynamic containers (`Grid`, `StackPanel`, and `WrapPanel`) all handle layout in this simple manner. The container in this example is the `Grid`. The `Grid` is different (and probably the most common of the containers) in that it allows you to create rows and columns. Blend makes it easy to create these rows and columns.

In a new simple Windows Phone application, Blend creates three `Grids` (`LayoutRoot`, `TitlePanel`, and `ContentPanel`). If you expand the `LayoutRoot` and select the `ContentGrid` in the Objects and Timeline panel, you will see that Blend shows a top- and left-side gutter for the grid, as shown in Figure 5.22.

These gutters are important as you can click on them to create columns and rows. Clicking on the grid will create a split in the container that

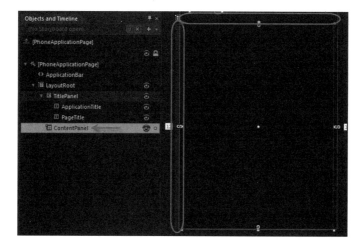

FIGURE 5.22 Column and row gutters

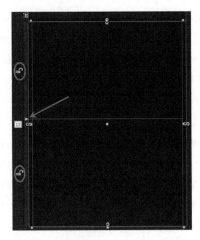

FIGURE 5.23 **Splitting the grid into rows**

indicates where two rows or columns are separated. For example, if you click on the left-hand gutter about halfway down you'll see two rows created, as shown in Figure 5.23.

Once the grid is split into two rows, you should notice that there are icons next to each row that indicate the way the row (or column) is sized. As you read in Chapter 3, XAML Overview, there are three types of sizing: auto, fixed, and star sizing. When you click on the icon, it will toggle through these three sizing types. Table 5.2 lists the icon for each sizing type.

When you click on the arrow indicator of the row (or column), the row above (or column to the left) is selected and the Properties panel will let you make changes manually, as shown in Figure 5.24.

When working with multiple columns and rows, you can stretch objects across the row and column boundaries. Blend will do its best to help you

TABLE 5.2 **Row/Column Sizing Icons**

Icon	Sizing Type
	Star
	Fixed
	Auto

Figure 5.24 Modifying row/column properties

determine which columns and rows you want to be a part of, but it is just guessing your intent. For example, you can see in Figure 5.25 that when an item just enters a second row it handles this with a negative margin value.

But if you stretch that more, Blend assumes you want to span the two rows, as shown in Figure 5.26.

You will often need to manipulate the Row, Column, RowSpan, ColumnSpan, Alignment, and Margin entries manually in the Properties panel to get the exact behavior you want. In many cases you may not want

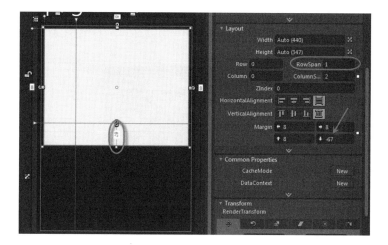

Figure 5.25 Sizing across rows

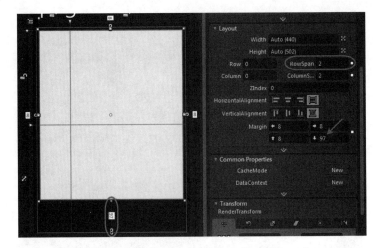

FIGURE 5.26 **Sizing across rows with** RowSpan

to use margins at all, but rather make your object a fixed size. You can do this by ensuring that your object is not set to Stretch alignment (vertically and horizontally). Then the Margin box will indicate where to draw your element, but not how to size your element.

Brushes

Another common task in Blend is to select how to "paint" a particular element of your design. As you saw in Chapter 4, Controls, Silverlight uses brushes to paint surfaces and Blend supports this through the Brushes section of the Properties panel, as shown in Figure 5.27.

The Brushes section of the Properties panel is made up of a number of sections you will want to become familiar with. The first section (#1 in

FIGURE 5.27 **Brushes in the Properties panel**

Figure 5.27) lists the brushes the particular object supports. In this example it's a simple `Rectangle` object, so you can set the brush for both the `Fill` and the `Stroke` of the `Rectangle`. This is like a small `ListBox`, so picking a brush will let you pick the properties for that brush in the lower part of the Brushes section. The second section (#2) lists the available brush types. This section takes the form of a set of tabs for each type of brush. Table 5.3 lists the brush editors.

TABLE 5.3 Brush Editors

Brush Type	Description	Editor
None	No brush will be drawn.	No editor is shown.
Solid color	Used to create solid color brushes. At #1 in the figure to the right there is a tab to allow you to design colors. The second tab is used to reuse resource-based colors. You can create resource-based colors by using the square context menu on the editor tab. At #2 is a simple color picker with a mix between white, black, and the selected tint. At #3 is where you can specify the red, green, blue, and alpha (transparency) manually. The bar at #4 allows you to pick the current tint that the mixer (#2) uses. The control at #5 allows you to specify or copy the specific color text. You can paste Web values here to match colors (e.g., from CSS). The last three elements are the current color (#6), the last color used (#7), and the eyedropper to pick a color (#8).	

continues

TABLE 5.3 Brush Editors (*continued*)

Brush Type	Description	Editor
Gradient	At #1 in the figure to the right, the gradient brush editor supports a bar that contains all the stops of a gradient brush. Clicking on the bar will create a new stop. The current stop is always drawn with a black background. When editing a stop on the gradient, the solid color editor (#2) is used to change the color at that stop. The only difference between the color editor and the gradient editor is that the simple eyedropper has been replaced with a gradient eyedropper (#3) to copy existing gradients. The editor rounds out with several controls on the bottom, including the switch between linear and gradient brushes (#4), the reverse stops button (#5), the stop navigator (#6), and the manual entry of a stop's offset (#7).	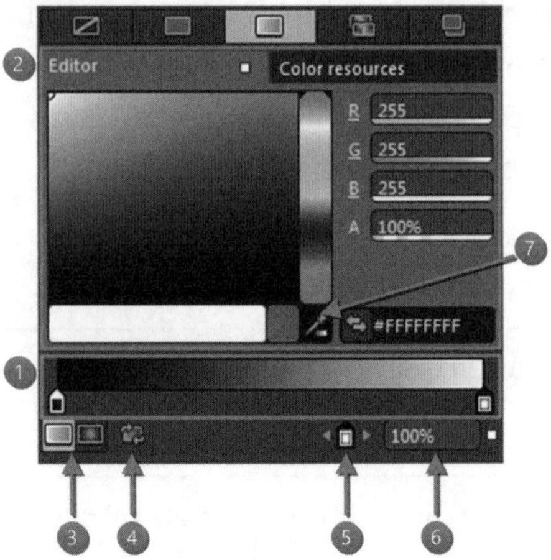
Image	The image brush editor allows you to specify the source of the image brush (#2 in the figure to the right) as well as how the image is fit to the source (e.g., Stretch) as seen in #1. There is also a preview of how the image will be stretched (#3).	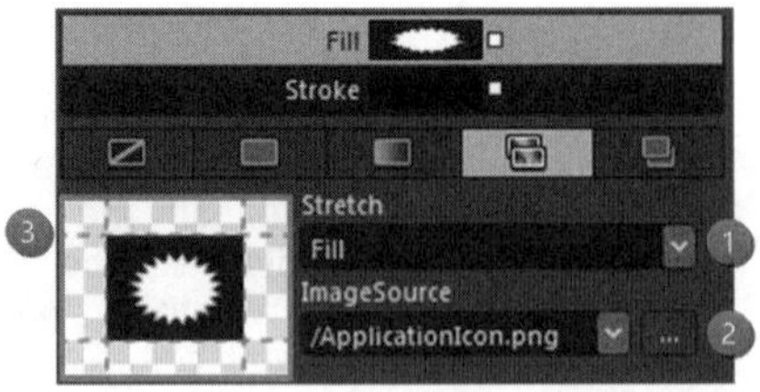

TABLE 5.3 Brush Editors (*continued*)

Brush Type	Description	Editor
Resource	The resource brush editor lists named brushes (both from your project and from the Windows Phone standard system brushes). You can simply pick from the list to pick a brush.	

Within the brush editor you can directly create two types of resources: color resources and brush resources. These are created in different ways, but both originate on the brush editor. First, let's create a color resource. We can do this by clicking on the small square next to the color picker, as shown in Figure 5.28.

Selecting Convert to New Resource will open a dialog where you can create a new named resource that will show up in the Resources list in the color editor, as shown in Figure 5.29.

Once you name the resource, you will also need to specify where to define the resource. Your options are labeled "Application," "This document," and "Resource dictionary." Typically you would choose "Application" or "Resource dictionary" if you have one. Defining a resource at the application level allows the resource to be used on any page in your application. Defining this resource allows you to specify it by name, and if it's

FIGURE 5.28 Converting a color to a resource

FIGURE 5.29 Creating a color resource

changed later in the design workflow, the change will affect every use of this color. This color resource can also be used in other brushes (e.g., gradient brushes). Once you have a color resource, you can use it via the "Color resources" tab, as shown in Figure 5.30.

The other type of resource is a brush resource. This is different as instead of defining a color to be used in a brush, you can define an entire brush (e.g., image, gradient, or solid color brush). This way, you can define a named brush so that if you later change a brush from a solid color brush to a gradient brush the change cascades to all uses. You can create a brush resource like you created a color resource, but the context menu is in a different location. This time you will use the small square to open the context menu on the brush itself instead of the color, as shown in Figure 5.31.

Applying the brush resource is similar, too. You would show the Resource Brush tab and pick the local brush resource you want to use, as shown in Figure 5.32.

Dealing with brushes is a pretty typical task. Using these techniques, you'll be up to speed in using Blend to create your Windows Phone applications quickly.

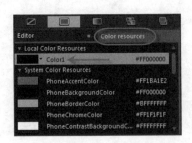

FIGURE 5.30 Applying a color resource

FIGURE 5.31 Creating a brush resource

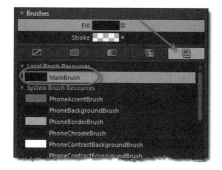

FIGURE 5.32 Applying a brush resource

Creating Animations

In Blend, the panel you use most often is the Objects and Timeline panel. It may not be obvious yet why it has this long name. The "Timeline" part of the name indicates that it is also where you can build your animations as part of your overall design. Timelines are the basic building blocks of animations.

To begin take a look at Figure 5.33. At the top of the Objects and Timeline panel is a bar that controls the creation of storyboard objects. A **storyboard** is a container that can hold animations of objects in your design. The + button in the bar allows you to create a new storyboard to contain animations.

FIGURE 5.33 Storyboard basics

FIGURE 5.34 Creating a storyboard

When you click the + button to start a storyboard, you're asked to name your new storyboard. This name will be used later to execute the storyboard (i.e., cause the animations to run), as shown in Figure 5.34.

Once you've created a storyboard, the Objects and Timeline panel shows a timeline next to each element in the object tree. Figure 5.35 shows the metamorphosis to this new look.

You can see and change which storyboard is the current one, using the panel below "Objects and Timeline" in Figure 5.35. The animation pane to the right shows the new storyboard of animations. The VCR-like controls at the top left corner of this pane allow you to play, pause, or move the current marker (the line currently positioned at 0) in the animation. You can also move the current marker by simply clicking at the top of the numbered timeline (these are in seconds).

To create your first animation you can simply click on the timeline to change the current time in the animation to one second. Also select the

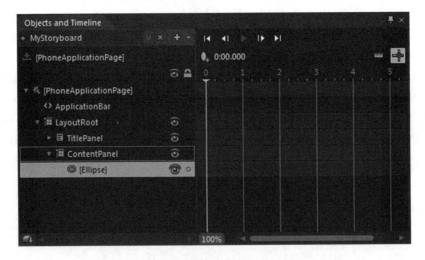

FIGURE 5.35 Objects and Timeline panel with animation

FIGURE 5.36 Picking the animation point

ellipse object in your object tree. Once you do that, you can change the properties of your ellipse and those changes will be reflected in the form of an animation. You can see how this looks in the Objects and Timeline panel, as shown in Figure 5.36.

At this point, you will notice that there is a red rectangle around the entire artboard to indicate that you are in animation mode and recording an animation (instead of just modifying objects normally). You can see this in Figure 5.37.

FIGURE 5.37 Animation mode on the artboard

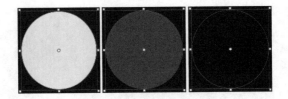

FIGURE 5.38 The ellipse animated

When you are in animation mode and you modify objects (through either the artboard or the Properties panel) your changes will be reflected in the storyboard. For this example, change the opacity of the ellipse to zero. This will cause the animation to move for one second (where our marker currently is set) and become transparent. You can see the state of the ellipse over the time of the animation (at zero, one-half, and one second) in Figure 5.38.

Back in the Objects and Timeline panel you can see that the ellipse now has a little red circle on it (to indicate it's part of this animation). It also shows an indicator at the one-second mark to show that the ellipse has a value at that part of the animation. If you open the arrow next to the ellipse, you will see that there is now a subobject that represents the property that was animated (e.g., opacity). If more than one property was affected, there would be a line for each of them. This is shown in Figure 5.39.

Instead of changing a property, you may decide to change the location or size of an object on the artboard. When you do this, Blend attempts to do what you want. To make it easy to animate, Blend usually incorporates

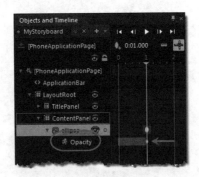

FIGURE 5.39 Animation values in the Objects and Timeline panel

FIGURE 5.40 `RenderTransform` **in an animation**

render transforms so that moving or sizing an element is based on the transform (instead of moving the margin and/or size of an element). If you move and resize your ellipse, you will see that Blend creates a `Render-Transform` and then creates values for the `RenderTransform`'s properties, as shown in Figure 5.40.

By adding these together, you can create complex animations, and then use the VCR-like controls to test them. Once you get the style you want, you can save and close the animation by clicking on the X button, as shown in Figure 5.41.

While you can create individual animations this way, you will find that using the visual state manager (often called the VSM) to create the state of a control or object is much more common. See the Blend documentation about the States panel for more information on the VSM.

Working with Behaviors

Although most of our discussion about design has focused on structure, some parts of a design should focus on performing actions. This is where behaviors come into play. A **behavior** is an object that can be activated based on an event in the design to perform some action. For example, you could have an animation fire when an item is acted upon. Figure 5.42 shows the different behaviors that ship with Blend.

Table 5.4 explains the different built-in behaviors.

FIGURE 5.41 **Closing a storyboard**

FIGURE 5.42 Behaviors in the Assets panel

TABLE 5.4 Blend Behaviors

Behavior	Description
ChangePropertyAction	Allows you to change a property value based on user input
ControlStoryboardAction	Allows you to change the state of an animation (usually start it)
FluidMoveBehavior	Animated changes to layout within a container to be smooth (or fluid)
FluidMoveSetTagBehavior	Same as FluidMoveBehavior but only affects objects with a specific Tag value
GoToStateAction	Controls a VisualStateManager .VisualStates to move to a specific state
MouseDragElementBehavior	Allows movement of an object when dragged by the user
NavigateToPageAction	Allows navigation to different pages within the Navigation Framework
PlaySoundAction	Causes a specific sound to be played when an action occurs
RemoveElementAction	Removes a specific child of a container when an action occurs

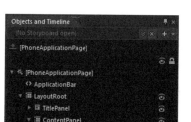

FIGURE 5.43 Applying a behavior

To use behaviors, you simply drag them onto an object in the Objects and Timeline panel or onto the artboard. For example, if you drag a ControlStoryboardAction object onto the ellipse it will create the new behavior on the ellipse, as shown in Figure 5.43.

Once the behavior is created, you'll see the behavior properties in the Properties panel. The top half (the trigger) is where you can pick the event that causes the behavior to fire. In this case we are executing it when the MouseLeftButtonDown event is fired, which also happens when a user presses the screen of Windows Phone. The top half is the same for most behaviors.

The bottom half is different for different behaviors. In this case we are choosing what to do to a storyboard (e.g., play, pause, etc.) and what storyboard to affect. The drop down will show you all the animations you've created (as shown in Figure 5.44).

You can have multiple behaviors on objects in the Objects and Timeline panel. In fact, you can have multiple behaviors with the same trigger. You can see the behaviors directly in the object tree, as shown in Figure 5.45.

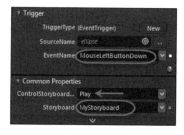

FIGURE 5.44 Changing behavior properties

FIGURE 5.45 Multiple behaviors

Behaviors do not replace the need for typical code, but they do represent a way to perform UI-specific operations (e.g., actions). For example, you might have a fly-out panel that is shown when a button is pressed. It is not necessary to get code involved to show and hide the panel as there is no logic in this operation. It's purely a UI operation. Behaviors allow you to have XAML that represents a unit of the UI. Anytime you put real logic (e.g., validation or other behavior) in the design, you're probably making a mistake and should do that sort of work in code.

Phone-Specific Design

Although designing an application using Blend should be similar whether you are building a traditional Silverlight application or a Windows Phone application, some things are only applicable to the phone. These include the `ApplicationBar`, the `Panorama` control, the `Pivot` control, and launching the emulator.

The `ApplicationBar` in Blend

While having a user interface that compels users to move, swipe, and pan their way to getting the most from your application is nice, sometimes users need simple operations. This is where the `ApplicationBar` comes in. The `ApplicationBar` is a standard part of many applications. A mix of a menu and a toolbar, the `ApplicationBar` can hold both icons (like a toolbar) and menu items. The `ApplicationBar` starts in a minimized (or closed) state and can be opened by the user to reveal more information, as shown in Figure 5.46.

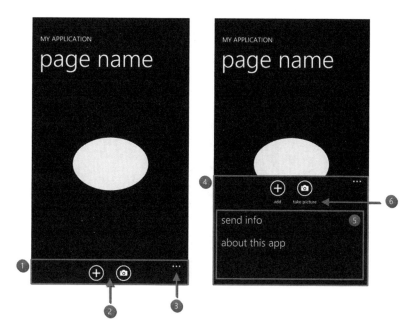

FIGURE 5.46 `ApplicationBar` **explained**

By default, the `ApplicationBar` takes up a small portion of the user interface (labeled #1 in Figure 5.46). On the `ApplicationBar` are icons that can perform certain tasks. In this mode only the icons are visible; the menu items are not. You can see the icons in the `ApplicationBar` (#2). To the right of the icons is an ellipse (#3) that alerts the user that pressing it will open the `ApplicationBar` (#4). In this open view, the menu items are shown below the icons (#5). Finally, you should notice that, when opened, the icons have text associated with them (this is not visible when closed, but is shown when the application bar is opened, as shown in #6). In this example, all the text is lowercase. The `ApplicationBar` forces any text in the menu items or icon text to be lowercase (to preserve the Metro-style guide).

You can add an `ApplicationBar` to a XAML file with the context menu on the `PhoneApplicationPage` element in the Objects and Timeline panel, as shown in Figure 5.47.

This adds the bar to your application as just another part of your object tree. From that point you can use the context menu to add both icons and menu items, as shown in Figure 5.48.

FIGURE 5.47 Adding an `ApplicationBar`

Adding an `ApplicationBarIconButton` creates the icons you see on the bar (whether closed or opened). Blend supports a number of built-in icons you can choose, or you can supply your own. With an `Application-BarIconButton` selected in the Objects and Timeline panel you can use the drop down to pick an icon to use for your application, as shown in Figure 5.49.

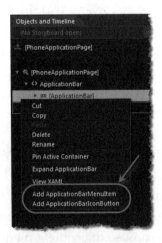

FIGURE 5.48 Adding items to the `ApplicationBar`

FIGURE 5.49 Selecting a built-in icon for an `ApplicationBar` **icon**

Selecting a built-in icon will add that icon to your project (in an Icons folder). You can simply supply your own `IconUri` to point to an icon of your choosing (located in your project). The built-in icons are useful, but be careful not to reuse a built-in icon for a use that is not natural. It will confuse users if you reuse an icon that is commonly used for some other function.

Typically these should follow the Metro style and be monochrome (e.g., white on black or black on white) in the PNG format. If you choose white with PNG transparency, Windows Phone will change your icons to black on white when the user switches his theme to a light theme. Otherwise, you will need to modify the icon at runtime manually.

Using the `Panorama` **Control in Blend**

On Windows Phone, you are not limited to the size of the screen. As we discussed earlier in this chapter, the `Panorama` control in an application allows you to use a larger virtual space. While you can use a `Panorama` control directly, one of the most common scenarios is to start with a new panorama application, as shown in Figure 5.50.

Once you create a new panorama application, your main page (Main-Page.xaml) will consist primarily of a virtual area that contains one or

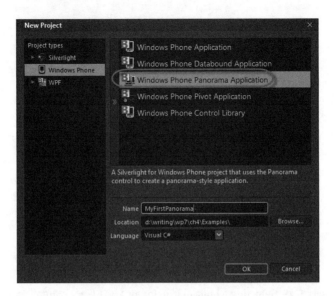

FIGURE 5.50 New panorama application

more `PanoramaItems`. We can see this if we drill down into the Objects and Timeline panel for the panorama, as shown in Figure 5.51.

The `Panorama` control user interface is made up of several parts, labeled 1–4 in Figure 5.52.

The control itself is sized to contain all panorama items. This means the control is much larger than the screen but indicates to the user that there is more content available (via the other panes overlapping on the side of the page), as indicated by the #1. The header (#2) shows the header for the current `PanoramaItem`. The #3 indicates the content area of an individual item. Lastly, the next item (#4) is shown to the right of the current item.

FIGURE 5.51 `PanoramaItems` in the Objects and Timeline panel

FIGURE 5.52 `Panorama` **control user interface**

This user experience is meant to help users learn your user interface by exploring the parts they can't see in the virtual space on the phone.

In Blend, you can view each `PanoramaItem` as the "current" item by simply selecting it in the Objects and Timeline panel, as shown in Figure 5.53.

As the individual `PanoramaItem` is selected, you can manually edit the contents as you would in any other container (e.g., `Grid`, `Canvas`, etc.). The actual `PanoramaItem` contains both a header and content. The

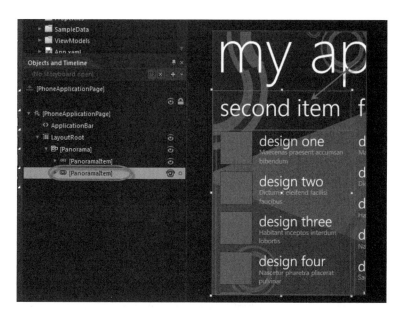

FIGURE 5.53 `PanoramaItem` **selection**

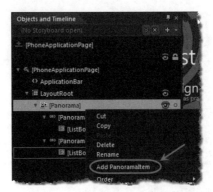

FIGURE 5.54 **Adding a** `PanoramaItem`

header (usually just lowercase text) indicates to the user what the `Pano-ramaItem` contains. The content is the control(s) that represents the content of the `PanoramaItem`.

Adding new `PanoramaItems` to your `Panorama` control is just as easy as using the context menu, as shown in Figure 5.54.

By adding multiple `PanoramaItem` objects to the `Panorama` control, you will be creating an increasingly larger virtual area for your application. As stated earlier, the rule of thumb is to only have four or five `Pano-ramaItems` in a `Panorama` control.

Using the `Pivot` Control in Blend

Much as you can with the `Panorama` control, you can use the `Pivot` control in an existing project or as the basis for a brand-new Windows Phone application. When creating a new project in Blend, you are given the option to create a new pivot application, as shown in Figure 5.55.

Once you create the pivot application, your main application is made up of a single `Pivot` control (much like the panorama above). The `Pivot` has `PivotItem` objects that represent a container for each pane in the `Pivot` control, as shown in Figure 5.56.

The user interface of the `Pivot` control is different from the `Panorama` control UI in that it is optimized for the screen, and navigating to the different items is accomplished via swipes or via the list of items at the top of the `Pivot` control. You can think of the `Pivot` as a view manager or even a tab control. The `Pivot` control UI consists of multiple parts, as shown in Figure 5.57.

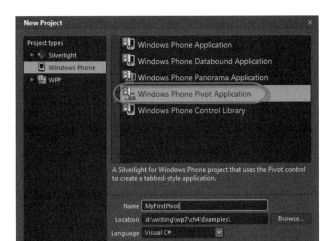

FIGURE 5.55 Creating a pivot application

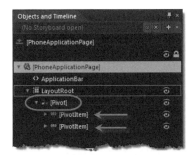

FIGURE 5.56 A pivot application

The section labeled #1 shows a header for each item of the `Pivot` control. Each item has a header (usually text) that the user can tap on to make that view the current view. The current view's header is shown on the left of this section and is typically highlighted (#2). Lastly, the main content of an individual item of the `Pivot` can contain any user interface that fits with the design (#3). Unlike the `Panorama`, the only indication that there is more information to view is the list of headers at the top.

Just like the `Panorama` control, you can edit the `Pivot` control's individual items (`PivotItems`) in Blend by simply selecting the appropriate `PivotItem`, which makes that item the current container in which you can do your design, as shown in Figure 5.58.

FIGURE 5.57 `Pivot` **control user interface**

At this point you can think of each `PivotItem` as just another container in your design.

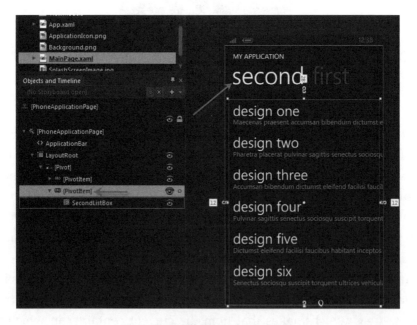

FIGURE 5.58 **Editing a** `PivotItem`

Previewing Applications

As you are working on the design of your application, you may want to test the look of the application. You can do this by using the Project menu (Project | Run Project). The Visual Studio hotkeys for this also match (F5 or Ctrl-F5) to launch your project. To facilitate previewing your application in a variety of scenarios, there is a Device panel in Blend that allows you to specify how to run your application. Figure 5.59 shows this panel.

The Device panel contains just three types of options. You can pick the orientation to choose when the application starts (#1), you can show the application theme (#2), or you can pick the theme and accept the color. The theme can be dark or light and the color is the chosen accent color to set on the device/emulator. Lastly, you can specify whether to preview your application via the emulator (the default) or on a device (#3). To test it on a device, you have to go through a couple of steps, including registering the phone with Microsoft as a development device. These steps are covered in Chapter 2, Writing Your First Phone Application.

Where Are We?

Design is an important part of any software project, but when you're writing for Windows Phone, you should treat design of the user experience as a crucial part of the process. Picking the right user paradigm is very important as the user cannot drop down into a richer set of input mechanisms (e.g., keyboard, mouse). Touch changes the way applications should work. In addition, applications should thrive on users' curiosity; compelling users to want to learn what your application is capable of is an important dynamic. Lastly, any design that does not make it clear from the start

FIGURE 5.59 Changing device properties

both what it is used for and how to use it is doomed to fail. When it comes to achieving your vision, Expression Blend is a great tool. It's important to realize that even if you're a developer, Blend is a *design* tool, not a *designer's* tool. Learning how to navigate it can really help you not only design great applications but also accomplish the task in a shorter time. Angle brackets are cool, but creating a complex design with a real tool (like Blend) lets you focus on the parts of the application authoring process you love; whether that is the user interface, the business logic, or the backend.

Though this chapter just scratched the surface of real design, hopefully it has given you a sense of what is required to get your vision onto the phone. You may want to augment this book with other design resources to elevate your applications to true experiences.

6

Developing for the Phone

T hus far in the book we have focused on building experiences. While creating the appropriate design for your application is crucial to being successful, the development process on the phone is what brings it all together. Developing your application for the phone requires that you understand the way Silverlight works on the phone. That is what you will learn in this chapter.

Application Lifecycle

When you create a new Silverlight for Windows Phone project, the project contains all the files you need to get started building a phone application. There are several key files, but let's start with the XAML files. As you can see in Figure 6.1, the App.xaml and MainPage.xaml files have code files associated with them (they have a ".cs" extension because this project is a C# project).

These code files represent the code that goes with the main page and the application class, respectively. As we discussed in Chapter 2, Writing Your First Phone Application, the App.xaml file is where we can store application-wide resources. The class that goes with the App.xaml file represents the application itself. In fact, when your application is started it is this file (not MainPage.xaml's class) that starts up first. This class is called the App class and it derives from the Application class:

FIGURE 6.1 Important files in a new project

```
public partial class App : Application
{
  // ...
}
```

This class is used to store application-wide code/data. For example, the App class stores the main frame for your entire application. This Root-Frame property exposes the frame in which all of your pages will display:

```
public partial class App : Application
{
  /// <summary>
  /// Provides easy access to the root frame
  /// of the Phone Application.
  /// </summary>
  /// <returns>The root frame of the Phone Application.</returns>
  public PhoneApplicationFrame RootFrame { get; private set; }

  // ...
}
```

While the App class represents your running application, it is exposed as a singleton via the Application class's Current property. For example, to get at the current application class anywhere in your code, you would call the static Current property on the App class:

```
Application theApplication = App.Current;
```

You should notice that the Current property returns an instance of the Application class (the base class). If you want to access properties on the App class instance itself, you must cast it to the App class, like so:

```
App theApplication = (App)App.Current;
var frame = theApplication.RootFrame;
```

```
// ...or...
frame = ((App)App.Current).RootFrame;
```

During initialization of the `App` class, the `RootFrame` is created and shown to the user of the phone. But if you look through the code for the `App` class it may not seem obvious how the MainPage.xaml is actually shown. The trick is in the third file I highlighted in Figure 6.1: WMApp-Manifest.xml.

The manifest file contains a variety of settings that help the phone (and the Marketplace) determine information about your application. One of the pieces of information in this file is how to start up the application. The Tasks section of the file contains a `DefaultTask` element that defines the `NavigationPage` to start your application. You can see that here:

```
<?xml version="1.0" encoding="utf-8"?>
<Deployment ... >
  <App ... >

    <!-- ... -->

    <Tasks>
      <DefaultTask Name ="_default"
                   NavigationPage="MainPage.xaml"/>
    </Tasks>

    <!-- ... -->

  </App>
</Deployment>
```

Once the `App` class is created and initialized, the application is told to navigate to the URI in the `Navigation` attribute of the `DefaultTask`. That is how your MainPage.xaml is shown.

Changing the Name of the XAML File

If you decide to change the name of the XAML file (and the underlying class file) you must be sure to change the name in the `x:Class` declaration as well.

Using a mix of the application class and the manifest file your application gracefully starts up with your first page being shown. This is similar to how running a typical desktop application starts as well, but the life-cycle of your phone application is actually much different from that.

Navigation

Your application does not have a window. This is an important indication that although it is called Windows Phone, the "Windows" in the name is not an indication that *Windows* is the operating system. You will have to get used to a different style of development. The entire application model is built around the concept of page-based navigation, which should be comfortable to any user of the Web.

When you start an application, it navigates to a particular page (as dictated in the WMAppManifest.xml file's default task). As the user navigates to other pages, the stack of pages grows. The user can press the Back button on the phone to return to the last page within your application. If the user is at the first page of your application, pressing the Back button will exit your application, as shown in Figure 6.2.

This page navigation implies a lesson to be learned that when writing Windows Phone applications there is no obvious way to exit an application. Instead of supporting a method to close the application, the Navigation Framework stipulates that when a user presses the Back button from the first page in your application it will exit as shown in Figure 6.2.

> ### ▪ For Silverlight Users
>
> If you have used the standard version of Silverlight, you will notice that the navigation API looks the same. The big difference is that not all navigation patterns are supported, even if the API makes it look like they are. For example, `GoForward` and `CanGoForward` exist on the `NavigationService` but are not supported.

Programmatically, you can interact with the navigation facility using the `NagivationService` class. To get access to the application's instance of the `NavigationService` class you have two options to find the navigation APIs.

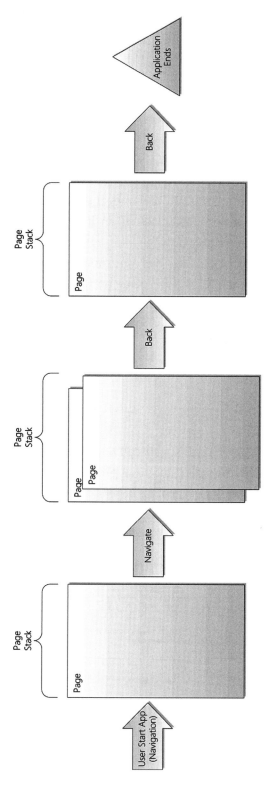

FIGURE 6.2 Page navigation explained

- `NavigationService` property on classes that derive from `Phone-ApplicationPage` (e.g., MainPage.xaml or other "Page" project items added to your project).
- `PhoneApplicationFrame` class, which has all the navigation functionality of the `NavigationService` class. This frame is ordinarily a member of your application's `App` class.

Although they are functionally the same, these two implementations do not depend on a common base class or an interface to enforce them. They are identical by convention only.

The `NavigationService` class has a number of useful methods, but the most common usage includes `Navigate` and `GoBack`. The `Navigate` method starts a navigation to a new "page," like so:

```
NavigationService.Navigate(new Uri("/Views/SecondPage.xaml",
                                   UriKind.Relative));

// or

NavigationService.Navigate(new Uri("/Views/SecondPage.xaml?id=123",
                                   UriKind.Relative));
```

The URI in the `Navigate` call specifically searches for a file in the .xap file and this is typically mimicked by your project structure, as shown in Figure 6.3.

`HyperlinkButtons` are automatically wired to use the `Navigation-Service`, so this works fine in your XAML:

```
<HyperlinkButton NavigateUri="/Views/SecondPage.xaml"
                 Content="Go to 2nd Page" />
```

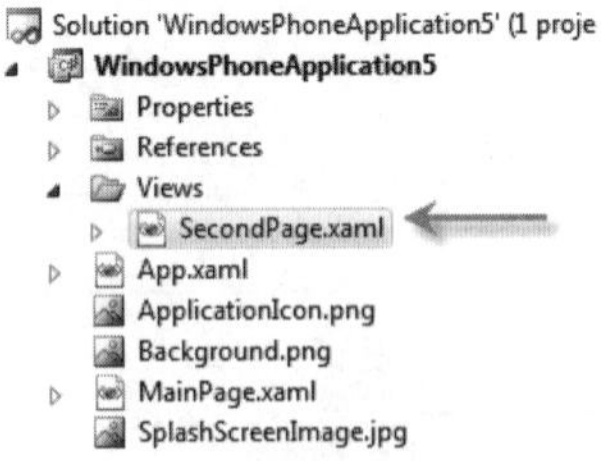

FIGURE **6.3** URI mapping to the files in the project

Conversely, the `NavigationService` class's `GoBack` method goes back to the top page on the navigation page stack (or back stack, explained in the next paragraph). This method mimics the user pressing the Back button:

```
NavigationService.GoBack();
```

As you navigate through an application, this `NavigationService` class keeps track of all the pages, so `GoBack` can walk through the pages as necessary. This stack of pages is called the **back stack.** The `Navigation-Service` provides read-only access to the back stack by providing a property:

```
IEnumerable<JournalEntry> backStack = NavigationService.BackStack;
```

The `backStack` property allows you to iterate through the back stack and interrogate the source of each page, but not change it:

```
// Iterate through the BackEntries
foreach (var entry in NavigationService.BackStack)
{
  Uri page = entry.Source;
}
```

The only change that the `NavigationService` allows is to remove the last entry in the `backStack`. You can do this with `RemoveBackEntry`, which removes only the current page from the `backStack`:

```
// Remove the last page from the navigation
NavigationService.RemoveBackEntry();
```

When navigation occurs (even at the start of an application), the class that represents the page has an opportunity to know it is being navigated to. This is implemented as overridable methods on the `PhoneApplica-tionPage` class. The first of these overridable methods is the `OnNavi-gatedTo` method:

```
public partial class MainPage : PhoneApplicationPage
{

  // ...

  // I was just navigated to
  protected override void OnNavigatedTo(NavigationEventArgs e)
  {
```

```
    base.OnNavigatedTo(e);

    var uri = e.Uri;

  }
}
```

Additionally, there are overridable methods for navigating away from a page (`OnNavigatingFrom` and `OnNavigatedFrom`):

```
public partial class MainPage : PhoneApplicationPage
{

  // …

  // Navigation to another page is about to happen
  protected override void OnNavigatingFrom(NavigatingCancelEventArgs e)
  {
    base.OnNavigatingFrom(e);
  }

  // Navigation to another page just finished happening
  protected override void OnNavigatedFrom(NavigationEventArgs e)
  {
    base.OnNavigatedFrom(e);
  }
}
```

The `NavigationService` also has events if you want to react to changes in the navigation as they happen, including `Navigating`, `Navigated`, and `NavigationFailed`. These events can be useful in globally monitoring navigation.

■ Accidental Circular Navigation

Be careful when doing navigation in your application as you can accidentally get into circular navigation by using the `Navigation-Service`'s `Navigate` method when you meant to use the `GoBack` method.

For example, you may want to go to an Options page in your application. When the user is finished changing options, it may be tempting to use `Navigate()` to return to the last page, but doing that leaves the last page in the page stack twice. Using `GoBack()` is the right way to return to the prior page.

As shown earlier, the URI you navigate to can contain query string information. You could get the URI in your `OnNavigatedFrom` method and try to parse the query string manually, but this is unnecessary. The `NavigationContext` class supports simple access to the query string. For example, if you wanted to get the ID from the query string, you could simply use the `NavigationContext` instead:

```
protected override void OnNavigatedTo(NavigationEventArgs e)
{
  base.OnNavigatedTo(e);

  if (NavigationContext.QueryString.ContainsKey("id"))
  {
    var id = NavigationContext.QueryString["id"];
    // Use the id
  }
}
```

By using the `NavigationService`, `NavigationContext`, and `PhoneApplicationPage` classes, you can control the navigation of different parts of your application as well as give users a more intuitive experience.

Tombstoning

As we saw back in Chapter 1, Introducing Windows Phone, the phone supports a concept called tombstoning. The idea of tombstoning is essentially to give the illusion of multitasking without the touch realities of running multiple applications at the same time. To achieve this illusion, your application can move between running, dormant, and suspended states, as shown in Figure 6.4.

When your application is on the screen (visible to the user) your application is in a running state. But a number of things can interrupt your application, including a phone call, a "toast" alert, or even the user pressing the Start or Search key. When something interrupts your application, your application is notified that it is being **deactivated.** During deactivation you can save certain data (or state) about your application. Once deactivation is complete, your process is moved to a **dormant** state. In this state all the threads are suspended and no processing continues to take place. If the operating system determines that it needs the memory your dormant application is using, it will remove it from memory (called the **suspended**

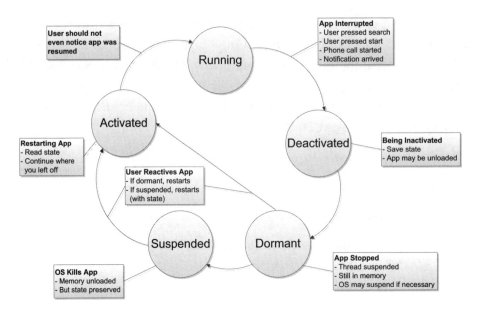

FIGURE 6.4 How tombstoning works

state), but will retain any data (state) you saved during deactivation. If the user returns to your app (usually via the Back button), your application is notified that it is being **activated.** During the activation process you will determine if the phone was activated from the dormant or suspended state and either just restart the application or use the saved state to return your application to its original state. As far as the user is concerned, the application should act just like it was running in the background.

■ Tombstoning

Tombstoning gives the "illusion of multitasking" without the overhead. It allows the operating system to load and unload your application as if it were running in the background. The user should be oblivious that your application was stopped or unloaded. It should "just work."

This process is exposed to you via the `PhoneApplicationService` class. An instance of this class is created in the App.xaml file of your project by default:

```
<Application ...>

  <!--Application Resources-->
  <Application.Resources>
  </Application.Resources>

  <Application.ApplicationLifetimeObjects>
    <shell:PhoneApplicationService
      Launching="Application_Launching"
      Closing="Application_Closing"
      Activated="Application_Activated"
      Deactivated="Application_Deactivated" />
  </Application.ApplicationLifetimeObjects>

</Application>
```

The `PhoneApplicationService` class is created to have the same lifetime as the `Application` class we discussed earlier. When it is created, it wires it up to four events in the `Application` class itself. The default Windows Phone Application template creates this object and wire-up for you. If you look at the App.xaml.cs/vb file (where the `Application` class is defined), you will see the corresponding methods that are mentioned in the `App.xaml` class:

```
public partial class App : Application
{
  // ...

  private void Application_Launching(object sender,
                                      LaunchingEventArgs e)
  {
    // ...
  }

  private void Application_Activated(object sender,
                                      ActivatedEventArgs e)
  {
    // ...
  }

  private void Application_Deactivated(object sender,
                                        DeactivatedEventArgs e)
  {
    // ...
  }

  private void Application_Closing(object sender,
                                    ClosingEventArgs e)
```

```
  {
    // ...
  }

  // ...
}
```

These boilerplate functions are called as the state of your application changes. The first and last ones (`Launching` and `Closing`) only happen once each: when your application is first launched directly by the user (e.g., by clicking on an application icon or responding to a toast notification) and when the user closes the application (e.g., by pressing back from the first page of the application). But for tombstoning, the real magic happens in the other two handlers: `Activated` and `Deactivated`. These handlers are called during tombstoning to allow you to save and load state into your application.

To save your state, the `PhoneApplicationService` class gives you access to a property bag to store data into. This class supports a `State` property, which is an `IDictionary<string, object>` collection that is used to store serializable objects by name. To use the class, simply use the class's static `Current` property to get at the current service object and use the `State` property from there. Here is an example of storing a simple object (a color) into the `State` property:

```
private void Application_Deactivated(object sender,
                                     DeactivatedEventArgs e)
{
    PhoneApplicationService.Current.State["favoriteColor"] = Colors.Red;
}
```

> ### ■ Note
>
> If you are new to Silverlight or .NET in general, the concept of serializable objects concerns whether an in-memory version of an object is compatible with being written out in a format for temporary or permanent storage. This storage is usually used to take objects in memory and store them for rehydration back to memory objects. Serialization is used when you're storing objects in the `PhoneApplication-Service`'s `State` collection, but it is also used when storing data to memory on the phone or even when you want to save data across a network connection.

The application is a common place to store this state, but since the `PhoneApplicationService` class is where the real magic happens, you can handle that state wherever you want. For example, in a single-page application, you might use the `PhoneApplicationService` class to store data as the page is navigated from (which happens during tombstoning as well):

```
protected override void OnNavigatingFrom(NavigatingCancelEventArgs e)
{
  base.OnNavigatingFrom(e);

  PhoneApplicationService.Current
                    .State[MoodKey] = moodPicker.SelectedItem;
}
```

Because the `PhoneApplicationService`'s `Current` method always points to the current service for the application, you can use this state class anywhere.

■ Saving Tombstone State

While tombstoning gives you the opportunity to save state, you should only save transitory state here. Tombstoning state should be just enough to restart the application. You can save longer-lived or non-volatile state into isolated storage (discussed in a later chapter) or in the cloud. The tombstoning state's lifetime is owned by the operating system and can go away without your control. Therefore, the tombstone state should only be for the minimum state required to have your application "wake up" from being suspended. The larger the tombstone state, the longer it will take you to wake up from tombstoning. Since the user shouldn't notice tombstoning, minimizing this state is important.

When your application is activated, you can check the event argument's `IsApplicationInstancePreserved` property to detect whether your application was dormant or actually suspended. If it was suspended you will need to recover the state saved during the `Deactivated` event:

```
private void Application_Activated(object sender,
                          ActivatedEventArgs e)
{
```

```
if (e.IsApplicationInstancePreserved)
{
   // Nothing to do as your application was just 'Dormant'
}
else
{
   // Recover State and resurrect your
   // application from being 'Suspended'
   var stateBag = PhoneApplicationService.Current.State;
   var state = stateBag["favoriteColor"];
   var color = (Color)state;
}
}
```

During activation you have a full ten seconds to process your data. Ten seconds ends up being a really long time and should not be used as the benchmark. Users will give up way before that. In general, tuning this to be as fast as possible will give you a better user experience.

The Navigation Framework is part of an application's tombstone state by default, so when an application is activated, the first page navigated to will be the last page the user was on. In addition, the page stack is also preserved, so you do not have to save this information manually. The applications will be activated in the exact same place (navigation-wise) from which they were deactivated. At this point you should understand how the life of a phone application works. Knowing that your application will need to deal with a mix of navigation, tombstoning, and startup/shutdown behaviors should equip you with the knowledge to build your application for the phone.

The Phone Experience

Phones are different. Yes, we've covered that already in this book, but it deserves to be reiterated. Phones are different. The development experience is different, too. While developing your application you will have to deal with several different facilities that relate to the user's experience on the phone: phone orientation, touch input, phone-specific controls, and the application bar. In this section you will learn how to deal with these different areas of functionality.

Orientation

The phone supports three main orientations: portrait, landscape left, and landscape right, as shown in Figures 6.5, 6.6, and 6.7. These orientations are supported on all the phones. You can control which orientation(s) your application supports as well as switching between them.

Figure 6.5 Portrait orientation

Figure 6.6 Landscape left orientation

FIGURE 6.7 **Landscape right orientation**

On the `PhoneApplicationPage` class, you can specify what orientations you expect your application to support:

```
<phone:PhoneApplicationPage
       ...
           SupportedOrientations="PortraitOrLandscape">
```

The `SupportedOrientations` accepts one of the values in the `SupportedPageOrientation` enumeration: `Landscape`, `Portrait`, or `PortraitOrLandscape`. By specifying `PortraitOrLandscape`, you are telling the phone that you want to allow the orientation to change as the phone is turned. By default, the phone will simply attempt to display your page using the new orientation. Since the default project template creates a page that is essentially just a grid within a grid, often this works. But in many cases you will want to customize the experience when the user changes the orientation. The `PhoneApplicationPage` class supports an `OrientationChanged` event that is fired when the orientation changes and this is a common way to change the user interface:

```
public partial class MainPage : PhoneApplicationPage
{
  // ...

  // Constructor
  public MainPage()
  {
    InitializeComponent();
```

```
    // Change size of app based on changed orientation
    OrientationChanged += MainPage_OrientationChanged;
  }
}
```

This will give you a chance to change your application as necessary. For example, if you just wanted to resize some of your user interface you could scale the application depending on the orientation:

```
void MainPage_OrientationChanged(object sender,
                               OrientationChangedEventArgs e)
{
  if ((e.Orientation & PageOrientation.Landscape) ==
                          PageOrientation.Landscape)
  {
    theScaler.ScaleX = theScaler.ScaleY = .86;
  }
  else if ((e.Orientation & PageOrientation.Portrait) ==
                          PageOrientation.Portrait)
  {
    theScaler.ScaleX = theScaler.ScaleY = 1.16;
  }
}
```

Although using a rendering transformation will help you change your user interface, you might decide to rearrange your design dramatically for the change in orientation, or even use a different view. The amount of change is completely up to you.

▪ Changing the Orientation Design in Blend

Jamie Rodriguez has a great presentation on some example behaviors he has created for Blend that can allow you to change the layout of your application using behaviors. See his video at http://channel9.msdn .com/blogs/jaime+rodriguez/windows-phone-design-days-blend.

Designing for Touch

The primary input on the phone is touch. It is the main reason that designing for the phone is so different from designing for other applications. At this point in the book you should have a good idea about the difference in the design metaphor, but programming against that metaphor is a different story.

If you are coming to Windows Phone development from Microsoft plat-forms (e.g., Silverlight, .NET, etc.), the easiest way to handle touch is to use the built-in mouse events. In fact, the touch interface is mimicked in the mouse events as you might expect. For example, to handle tapping on the surface of an application, you can simply handle the mouse event:

```
public partial class MainPage : PhoneApplicationPage
{
  // Constructor
  public MainPage()
  {
    InitializeComponent();

    ContentPanel.MouseLeftButtonUp +=
      new MouseButtonEventHandler(ContentPanel_MouseLeftButtonUp);

  }

  void ContentPanel_MouseLeftButtonUp(object sender,
                                      MouseButtonEventArgs e)
  {
    theText.Text = "Content Panel Tapped";
  }

  // ...
}
```

While the built-in mouse events do work, they are not used often as touch is very different from mouse input. We tend to drag, swipe, and pinch the screen very differently than we did (or could) with the mouse. The phone supports four points of touch so that the mouse events end up falling down on any input that allows more than one finger.

To help with touch, the phone has several layers of APIs to help you get to the touch surface. At the lowest level, the Touch class can report every touch interaction the user does. Silverlight's Touch class is most use-ful when you want to get as close to the metal as possible. The Touch class has a static event called FrameReported, which is called as touch interac-tion is happening. With it you can get information about how many touch points are being used (e.g., how many fingers are on the touch surface) as well as where the touch points are. The event contains an argument that allows you to get at the point information. Here is an example using the

`FrameReported` event to show where the touch is being dragged on the surface of the phone:

```
public partial class MainPage : PhoneApplicationPage
{
  // Constructor
  public MainPage()
  {
    InitializeComponent();

    Touch.FrameReported += new
      TouchFrameEventHandler(Touch_FrameReported);
  }

  void Touch_FrameReported(object sender, TouchFrameEventArgs e)
  {
    var mainTouchPoint = e.GetPrimaryTouchPoint(this);
    if (mainTouchPoint.Action == TouchAction.Move)
    {
      theText.Text = string.Concat("Moving: ",
                                mainTouchPoint.Position);
    }
  }

  // ...
}
```

The `TouchFrameEventArgs` class has several pieces of functionality. The two that are most important are the `GetPrimaryTouchPoint` and `GetTouchPoints` methods. The `GetPrimaryTouchPoint` method is used to retrieve a `TouchPoint` object that is relative to a particular `UIElement` of the design. The primary touch point is determined by the first thing that touches the screen. Being **relative** means that all touch positions will be relative to that `UIElement`. The `TouchPoint` class can tell you the position and action that are occurring, as shown in the following code sample:

```
void Touch_FrameReported(object sender, TouchFrameEventArgs e)
{
  // Get the main touch point (relative to a UIElement)
  TouchPoint mainTouchPoint = e.GetPrimaryTouchPoint(ContentPanel);

  // Get the position
  Point position = mainTouchPoint.Position;
```

```
// Get the Action of the Touch
switch (mainTouchPoint.Action)
{
  case TouchAction.Move:
    theText.Text = "Moving";
    break;
  case TouchAction.Up:
    theText.Text = "Touch Ended";
    break;
  case TouchAction.Down:
    theText.Text = "Touch Started";
    break;
}
}
```

The `GetTouchPoints` method also retrieves touch information that is relative to a `UIElement`, but in this case all the current touch points are returned as a collection of `TouchPoint` objects:

```
void Touch_FrameReported(object sender, TouchFrameEventArgs e)
{

  // Get all the touch points
  TouchPointCollection points = e.GetTouchPoints(ContentPanel);

  theText.Text = string.Concat("#/Touch Points: ", points.Count);

}
```

Each `TouchPoint` object in the collection represents a single touch point on the phone. All phones support at least four points of touch. The way you work with the individual touch points from the collection is identical to the `TouchPoint` you retrieved from the `GetPrimaryTouch-Point` method.

While the `Touch.FrameReported` event will give you a lot of control, you may want something higher-level so that you can handle simple movement or sizing behavior. To fill that need, Silverlight also supports manipulations. The concept behind manipulations is to be able to easily interact with common manipulations of objects on the screen. These include sizing and moving of objects. The `UIElement` class supports these directly by supporting three events, as described in Table 6.1.

These events are used to support manipulation of objects on the screen, more than just touch. The basic idea of a manipulation is to be notified

TABLE **6.1** Manipulation Events

Event	Description
ManipulationStarted	Occurs when a manipulation (pinch or drag) begins on the UIElement
ManipulationCompleted	Occurs when a manipulation (pinch or drag) is complete on the UIElement
ManipulationDelta	Fires as a manipulation (pinch or drag) is occurring on the UIElement

about attempts to change objects on the phone by dragging or resizing. For example, the ManipulationDelta event sends information about the manipulation while it is happening in the form of a ManipulationDeltaEventArgs object argument. This argument includes both the cumulative amount of manipulation and the difference between the last delta and the current one. The manipulation amount is defined in a class called ManipulationDelta. The ManipulationDelta class contains two pieces of information: translation and scale. These pieces of information correlate directly to the idea of how transforms in Silverlight work (e.g., TranslateTransform and ScaleTransform specifically). The amount of translation indicates how far an item has been dragged (or moved). The scale indicates how much the item has been sized using the pinch touch gesture. For example, to move an object using a manipulation you might add a TranslateTransform to your design:

```
...
<Ellipse Fill="Red"
         Width="200"
         Height="200"
         x:Name="theCircle">
  <Ellipse.RenderTransform>
    <TranslateTransform x:Name="theTransform" />
  </Ellipse.RenderTransform>
</Ellipse>
...
```

With the transform in place, you can handle the ManipulationDelta event and use the TranslateTransform to move the element in response to dragging:

```
public partial class MainPage : PhoneApplicationPage
{
  // Constructor
  public MainPage()
  {
    InitializeComponent();

    ManipulationDelta += new EventHandler<ManipulationDeltaEventArgs>(
      MainPage_ManipulationDelta);
  }

  void MainPage_ManipulationDelta(object sender,
                          ManipulationDeltaEventArgs e)
  {

    // Move (e.g. Translate) the ellipse based on delta
    ManipulationDelta m = e.CumulativeManipulation;
    theTransform.X = m.Translation.X;
    theTransform.Y= m.Translation.Y;

  }

  ...

}
```

In this particular example, the CumulativeManipulation is used to get the entire touch manipulation. The Manipulation's Translation property contains the amount of the translation, but we cannot be sure this manipulation is about a particular element on our page (as we registered for the page's ManipulationDelta event). We could register just for our ellipse's manipulation, but alternatively we could also test to see what container was being manipulated by testing the ManipulationContainer like so:

```
void MainPage_ManipulationDelta(object sender,
                            ManipulationDeltaEventArgs e)
{
  if (e.ManipulationContainer == theCircle)
  {
    // Move (e.g. Translate) the ellipse based on delta
    ManipulationDelta m = e.CumulativeManipulation;
    theTransform.X = m.Translation.X;
    theTransform.Y = m.Translation.Y;
  }

}
```

Using pinch and zoom touch gestures works in the same way you can use a `ScaleTransform` to change the size instead of the `TranslateTransform`:

```xml
<Ellipse Fill="Red"
         Width="200"
         Height="200"
         x:Name="theCircle">
  <Ellipse.RenderTransform>
    <ScaleTransform x:Name="theTransform"
                    CenterX="100"
                    CenterY="100"/>
  </Ellipse.RenderTransform>
</Ellipse>
```

Then, in the manipulation events, you can simply use the scale properties in the `ManipulationDelta` instead of `Translate`, like so:

```csharp
void MainPage_ManipulationDelta(object sender,
ManipulationDeltaEventArgs e)
{
  if (e.ManipulationContainer == theCircle)
  {
    // Size (e.g. Scale) the ellipse based on delta
    ManipulationDelta m = e.CumulativeManipulation;
    theTransform.ScaleX = m.Scale.X;
    theTransform.ScaleY = m.Scale.Y;
  }
}
```

Manipulations also include information about the **inertia** of the touch gestures. The idea behind inertia is to be able to tell if the user was still moving when the manipulation ended. The inertia information becomes important to making more organic interactions with users. If you have seen how flicking a list box on the phone makes the list scroll up even when the user is no longer touching the phone, this is accomplished using inertia.

For example, on the `ManipulationCompleted` event you can test for `IsInertial` to see if the manipulation contains inertial velocity information:

```csharp
void MainPage_ManipulationCompleted(object sender,
                                    ManipulationCompletedEventArgs e)
{
  if (e.IsInertial)
  {
    var m = e.TotalManipulation;
```

```
    var velocity = e.FinalVelocities;
    theTransform.X =
        m.Translation.X + (velocity.LinearVelocity.X / 100);
    theTransform.Y =
        m.Translation.Y + (velocity.LinearVelocity.Y / 100);
    }
}
```

Once the code determines it is inertial, it can use the FinalVeloci-
ties to change the outcome of the translation (in this example). You could
also see if the LinearVelocity is greater than some threshold to determine if it is a "flick" or not:

```
void MainPage_ManipulationCompleted(object sender,
                                    ManipulationCompletedEventArgs e)
{
    if (e.IsInertial)
    {
        var velocity = e.FinalVelocities;

        // Is it a Right Flick?
        if (velocity.LinearVelocity.X > 100)
        {
            // ...
        }
    }
}
```

The manipulation events are specifically to handle the drag and scale,
but as we discussed before, there are a number of types of touch gestures.
While having access to manipulations and lower-level access with the
Touch class helps, for most of your touch interface you'd like to access
events for those gestures directly.

The UIElement class has access to the most common types of touch
gestures. You can see the touch events that the UIElement class exposes
in Table 6.2.

Wiring up to these events is as simple as wiring up the event:

```
public partial class MainPage : PhoneApplicationPage
{
    // Constructor
    public MainPage()
    {
        InitializeComponent();
```

TABLE 6.2 `UIElement` Touch Events

Event	Description
`Tap`	Occurs when a user touches a `UIElement` and lifts her finger fairly quickly
`Double-tap`	Occurs when a user taps a `UIElement` twice in quick succession
`Hold`	Occurs when a user touches and holds her finger over a `UIElement`

```
    var listener = theCircle.Hold +=
      new EventHandler<GestureEventArgs>(theCircle_Hold);
  }

  void theCircle_Hold(object sender, GestureEventArgs e)
  {
    // Do a hold
  }
}
```

The `UIElement` class represents any visual element on a page, so you can wire up these events on any object (e.g., `Button`, `ListBox`, `Grid`, `Ellipse`, etc.). As you work with touch, you will use a variety of these touch-based APIs in your application. When working with common gestures, the element approach is easiest, but as you want more control over the nature of the touch surface you will have to delve further down into the stack of APIs.

Application Client Area

As the basis for the entire application, the shell is responsible for hosting your pages. This also means certain responsibilities are given to a set of classes that exist in the `Microsoft.Phone.Shell` namespace and it is where you can make some decisions about how your application is displayed. As explained in Chapter 1, the application area is broken up into a system tray, the logical client area, and the application bar (as shown in Figure 6.8).

FIGURE 6.8 Application client area

You can decide to hide the system tray (entering a sort of "full-screen mode") via the `SystemTray` class, as shown below:

```
void MainPage_Loaded(object sender, RoutedEventArgs e)
{
    SystemTray.IsVisible = false;
}
```

Using the `SystemTray` class enables you to change this to meet your needs. You do not have to set this in code, but doing so is supported via an attached property in XAML:

```
<phone:PhoneApplicationPage ...
  xmlns:shell=
    "clr-namespace:Microsoft.Phone.Shell;assembly=Microsoft.Phone"
  shell:SystemTray.IsVisible="True">
```

There is no actual "full-screen mode" as the Windows Phone documentation suggests; rather, this indicates whether the system tray is shown or not. Showing the system tray reduces the size of your application; it does not simply show "over" your application.

Several operations on the phone also will show over some part of your application (but not navigate away from your application). Receiving a phone call, alarms, and the lock screen are common examples of this. When the phone shows these over your application, it is **obscuring** your application. You can react to these activities by handling the Phone-ApplicationFrame's Obscured and Unobscured events:

```
var rootFrame = ((App)App.Current).RootFrame;

rootFrame.Obscured += new
  EventHandler<ObscuredEventArgs>(RootFrame_Obscured);

rootFrame.Unobscured +=
  new EventHandler(RootFrame_Unobscured);
```

These events will allow you to react to the obscuring of your application. A common usage is to enter a pause screen when you are writing a Silverlight game for the phone.

Application Bar

Another part of the normal screen real estate is the application bar. This bar is a mix of a toolbar and a menu. While you can create an application bar using Blend (i.e., XAML), programming against application bars presents a couple of small inconveniences. As I explained before, the Application-Bar supports ApplicationBarIconButton and ApplicationBar-MenuItem objects to allow users to interact with the application:

```
<phone:PhoneApplicationPage.ApplicationBar>
  <shell:ApplicationBar IsVisible="True"
                        IsMenuEnabled="True">
    <shell:ApplicationBarIconButton
      IconUri="/icons/appbar.add.rest.png"
      Text="add" />
    <shell:ApplicationBarIconButton
      IconUri="/icons/appbar.back.rest.png"
      Text="revert" />
    <shell:ApplicationBar.MenuItems>
      <shell:ApplicationBarMenuItem Text="options" />
      <shell:ApplicationBarMenuItem Text="about" />
    </shell:ApplicationBar.MenuItems>
  </shell:ApplicationBar>
</phone:PhoneApplicationPage.ApplicationBar>
```

Wiring up the Click events in XAML is possible and works the way you expect it to:

```
<shell:ApplicationBarIconButton
  IconUri="/icons/appbar.add.rest.png"
  x:Name="addIconButton"
  Click="addIconButton_Click"
  Text="add" />
```

But the application bar is different from typical buttons or similar Silverlight controls. The menu and icon button objects don't support commanding (like `ButtonBase`-derived controls), so you can't use data binding with the commanding support in Silverlight. Even if you name elements in the `ApplicationBar` (like above) the resultant wired-up member of the code-behind is null, so the event wiring will fail:

```
public partial class MainPage : PhoneApplicationPage
{
  // Constructor
  public MainPage()
  {
    InitializeComponent();

    // Doesn't work because addIconButton is null!
    addIconButton.Click += new EventHandler(addIconButton_Click);
  }
  ...
}
```

There is a technical reason this does not work,[1] but more important than understanding the reason is to understand that when you need to refer to the icon buttons or menu items in code, you will need to retrieve them by ordinal from the `ApplicationBar` property on the `PhoneApplicationPhone` class. While this is fragile and painful, it's the only way this will work currently. Here is how to do it:

```
public partial class MainPage : PhoneApplicationPage
{
  // Constructor
  public MainPage()
  {
    InitializeComponent();
```

---

1. The `ApplicationBar` is applied to the `PhoneApplicationPage` class via an attached property. This means the named elements are not part of the namescope, so calling the `FrameworkElement` class's `FindName` method that the generated partial class uses does not find the icon buttons or menu items.

```
    // Wire-up the button by ordinal
    addIconButton =
      (ApplicationBarIconButton)ApplicationBar.Buttons[0];

    // Wire-up the menu by ordinal
    optionMenuItem =
      (ApplicationBarMenuItem)ApplicationBar.MenuItems[0];

    // Works now that the manual wire-up was added
    addIconButton.Click += new EventHandler(addIconButton_Click);
  }
...
}
```

Understanding Idle Detection

The Windows Phone operating system automatically detects when a user has stopped using the phone and locks the phone. The phone will go to a lock screen. This lock screen can either display a pass code to open the lock screen or just instruct the user to slide up the wallpaper screen to get back to the phone.

Sometimes when you're building certain types of applications you want your application to continue to run regardless of whether the phone has input or not. There are two types of idle detection modes: application and user.

The easier of these to understand is the **user idle detection mode.** On the `PhoneApplicationService` class is a property called `UserIdle-DetectionMode` and it is enabled by default. This means that if no touch events are detected, it will let the phone go to the lock screen. You can change this behavior (to allow your application to continue to run as the main running application even if the user isn't interacting with your application) by disabling this detection mode like so:

```
// Allow application to stay in the foreground
// even if the user isn't interacting with the application
PhoneApplicationService.Current.UserIdleDetectionMode =
  IdleDetectionMode.Disabled;
```

Disabling the user idle detection mode is useful when you are doing things that engage the user but do not require input (e.g., showing a video, displaying a clock, etc.).

In contrast, the **application idle detection mode** is not about preventing the lock screen but determining what the application should do when the lock screen is enabled. By default, the application idle detection mode is enabled, which means that when the lock screen appears the application is paused (as though it was being tombstoned). By disabling the application idle detection mode, you allow your application to continue to run under the lock screen. To disable the application idle detection mode, you also use the `PhoneApplicationService` class:

```
// Allow your application to continue to run
// when the lock screen is shown
PhoneApplicationService.Current.ApplicationIdleDetectionMode =
  IdleDetectionMode.Disabled;
```

The Tilt Effect

In many places on the phone, a subtle but effective feedback mechanism is employed when selecting items in a list or other controls. This is called the **tilt effect.** In Figure 6.9, you can see that the second item in the list is not being interacted with touch; however, in Figure 6.10 you can see that the second item is subtly tilted to give the user feedback that she is touching the item. In fact, it's hard to see here in static images, but the tilt actually interacts with where the user is touching the item.

Unfortunately this effect is not built into the framework, but it is available in the Silverlight for Windows Phone Toolkit. When you include the

First ListBoxItem
Second ListBoxItem
Third ListBoxItem
Fourth ListBoxItem

FIGURE 6.9 Untilted

First ListBoxItem
Second ListBoxItem
Third ListBoxItem
Fourth ListBoxItem

FIGURE 6.10 Tilted

toolkit's XML namespace declaration in your XAML file, you can specify it at any level to enable or disable this behavior. The namespace required is the same one that is used to include Silverlight for Windows Phone Toolkit controls (as shown in Chapter 5, Designing for the Phone). Typically you would include it at the page level to support the tilt effect in all controls, like so:

```
<phone:PhoneApplicationPage ...
    xmlns:toolkit="clr-namespace:Microsoft.Phone.Controls;
                    assembly=Microsoft.Phone.Controls.Toolkit"
    toolkit:TiltEffect.IsTiltEnabled="True">
```

The `TiltEffect.IsTiltEnabled` property can be applied to any individual control as well if you want to apply the effect to specific controls instead of at the page or container levels. The other attached property is `TiltEffect.SuppressTilt`. This attached property allows you to turn the tilt effect off on particular controls where the `TiltEffect.IsTiltEnabled` attached property is enabled at a higher level. For example:

```
<Grid x:Name="ContentPanel"
      Grid.Row="1"
      Margin="12,0,12,0"
      toolkit:TiltEffect.IsTiltEnabled="False">
  <ListBox FontSize="28"
           Name="listBox1"
           toolkit:TiltEffect.SuppressTilt="True"
           ItemsSource="{Binding}" />
</Grid>
```

Because these are attached properties, you can set them in code as well:

```
using Microsoft.Phone.Controls;

public partial class MainPage : PhoneApplicationPage
{
  // Constructor
  public MainPage()
  {
    InitializeComponent();

    listBox1.SetValue(TiltEffect.SuppressTiltProperty, false);
  }
}
```

Where Are We?

At this point in the book you have only just begun to understand the basics of how to write code for the phone, but you should be comfortable with the lifecycle of phone applications as well as simple tasks for interacting with the user with touch and events. This chapter focused on showing you the fundamentals of developing Silverlight for the phone, including the differences in how Silverlight applications work on the phone (versus the desktop or Web applications written for Silverlight). By applying the basics of tombstoning, navigation, and touch interactions you should be able to build compelling user interfaces at this point.

7

Phone Integration

Building applications for Windows Phone OS 7.1 is a compelling experience. So far we have seen ways to bring the desktop or Web experience to a different form factor: the phone. But the phone has a number of device-specific features that you will need to become familiar with if you are to succeed in building compelling phone applications.

Using Vibration

Unlike computers, phones are held in the hand. This makes vibration a useful mechanism to alert users that something is happening with their phones. This is called **haptic feedback.** While haptic feedback is really useful in some scenarios, its overuse is discouraged. Use of vibration can impact battery life, so vibrating the phone with every button click or other action isn't recommended. In general, using vibration is recommended for actions for which it may be difficult to use a visual cue. Small touch points are a common place for this as the user cannot see the visual cue because his finger is often in the way.

Combining vibration with visual cues can really round out the user experience. To use haptic feedback, Windows Phone provides a simple class called `VibrateController` (in the `Microsoft.Devices` namespace):

```
private void theButton_Click(object sender, RoutedEventArgs e)
{
    VibrateController.Default.Start(TimeSpan.FromMilliseconds(10));
}
```

As you can see, the `VibrateController` class provides a static property to access the default controller. From there you can start or stop the vibration. Typically you would just call `Start` with a short amount of time, to give the user that haptic feedback. You can also use the `Stop` method to cancel a long vibration, but in almost all cases you're just creating a very short vibration for the user to know he performed some action in your application.

> ■ **Emulator Tip**
>
> `VibrateController` will run in the emulator but you won't be able to tell it is working. That means the window won't vibrate—really!

Using Motion

Every Windows Phone also has an accelerometer built in. The **accelerometer** is a sensor that helps determine the phone's speed and direction based on its relationship to gravity. This means you can determine not only its position in three dimensions (which is how the phone determines when to change orientation) but also how much force is applied in each direction. As a result, you can determine the direction as well as the force in that direction. This is how some applications can test for shaking or other movements of the phone. Phones starting with the 7.1 version of Windows Phone also have a gyroscope to increase the sensitivity of this functionality.

The accelerometer works by showing the force against gravity. The force is separated into three axes to allow you to determine the location and force on the phone based on its relative position to gravity. These axes (x, y, and z) relate to the phone's position, and each will typically be in the range of -1 to +1 based on its position. If force is applied to the phone, these ranges can increase to detect the amount of force. For example, when you shake the phone the range will typically be greater than 1 or less than -1 to reflect that you are applying a force to the phone greater than that of gravity.

You can use this information to determine the amount of tilt applied to the phone. You can determine the tilt by comparing the value of each axis. These axes are mapped to the phone itself, as shown in Figure 7.1 (using the emulator).

Each axis has a negative and positive direction. For example, when you are holding the phone in portrait mode, exactly level with the ground, the y-axis will be -1 to represent that the top of the phone is up. In that case, the x-axis will be zero as it is halfway between lying down horizontally in either direction. This is the case with the z-axis as well, since the front and back of the phone are neither up nor down. Conversely, if you lay the phone (faceup) on a perfectly flat table, the z-axis will be -1 and the other axes will be zero. You can use the emulator to emulate moving the phone in three-dimensional space to see how it can be affected.

To use the accelerometer API, you first need a reference to the `Microsoft.Device.Sensors.dll` assembly (in the SDK) as well as the main XNA assembly (`Microsoft.Xna.Framework`). The former assembly gives you access to the `Accelerometer` class and the latter gives you access to the `Vector3` class, which is used to communicate the

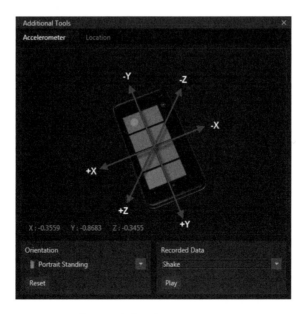

FIGURE 7.1 Accelerometer axes

three axes. To use the class you create an instance of it and register for the
CurrentValueChanged event, like so:

```
public partial class MainPage : PhoneApplicationPage
{
  Accelerometer _theAccelerometer = new Accelerometer();

  // Constructor
  public MainPage()
  {
    InitializeComponent();

    _theAccelerometer. CurrentValueChanged +=
      _theAccelerometer_CurrentValueChanged;
  }
```

You will also want to determine how often you want updates. This
allows the accelerometer to notify you only as often as necessary for your
application. For example, if you're writing a tilt-based game, updates
may be frequent (e.g., 100ms range). You can set this by setting the Time-
BetweenUpdates property of the Accelerometer class:

```
  // Constructor
  public MainPage()
  {
    InitializeComponent();

    _theAccelerometer.CurrentValueChanged +=
      _theAccelerometer_CurrentValueChanged;

    _theAccelerometer.TimeBetweenUpdates =
      TimeSpan.FromSeconds(1);
  }
```

The Accelerometer class supports two methods for starting and
stopping the accelerometer (not coincidentally called Start and Stop).
You will only want to enable the accelerometer when you actually need it.
Deferring its use until the user needs it is fairly typical. For example, you
might enable and disable it via buttons:

```
  void startButton_Click(object sender, EventArgs e)
  {
    _acc.Start();
  }
```

```
void stopButton_Click(object sender, EventArgs e)
{
  _acc.Stop();
}
```

Once you start the accelerometer, you will be notified as the readings change:

```
void _theAccelerometer_CurrentValueChanged(object sender,
                    SensorReadingEventArgs<AccelerometerReading> e)
{
  var position = e.SensorReading.Acceleration;

  // Update the User Interface
  Dispatcher.BeginInvoke(() =>
  {
    xValue.Text = position.X.ToString("0.00");
    yValue.Text = position.Y.ToString("0.00");
    zValue.Text = position.Z.ToString("0.00");
  });
}
```

The event argument (SensorReadingEventArgs<Accelerometer-Reading>) will pass you two pieces of information: a timestamp of when the reading changed, and the sensor reading itself (in the form of an instance of the SensorReading class). Inside the SensorReading's Acceleration property will be the X, Y, and Z values of the reading. Because the accelerometer can call you very quickly, the calls to the event do not happen on the UI thread (so the readings don't overwhelm the user interface thread). If you want to update the UI (like this example shows), you should marshal those calls to the UI thread (which Dispatcher. BeginInvoke does nicely for you).

▪ Tip

The accelerometer range is device-dependent. Don't depend on a single device to determine what the acceptable range is.

Emulating Motion

When developing an application that takes advantage of the accelerometer you can debug directly on a real device if that makes the most sense.

Otherwise, you can use the emulator to emulate motion for your development. The emulator has a button on the control bar to show a separate window with support for motion and location emulation. To show this additional window, you have to click the right-arrow icon on the control bar of the emulator, as shown in Figure 7.2.

Clicking on that button will open a window with an Accelerometer tab and a Location tab. To emulate motion, you will use the Accelerometer tab, as shown in Figure 7.3.

You should notice the image of a phone floating in space on the Accelerometer tab. In the middle of the phone is a pink circle that you can grab with your mouse to move it in 3D space. As you do that, the accelerometer values change (shown as X, Y, and Z in the lower left). This will allow you to see the relative axis values as you move the phone.

On the bottom-left area of the window you will see an Orientation drop down where you can change your perspective to the phone. For example, if you change the orientation to Portrait Flat, you'll still see the phone, but it will be from the perspective of the phone lying flat on a table.

Finally, on the bottom-right area of the window is the Recorded Data drop down that contains prerecorded sets of accelerometer data that you

FIGURE 7.2 Showing the Accelerometer window in the emulator

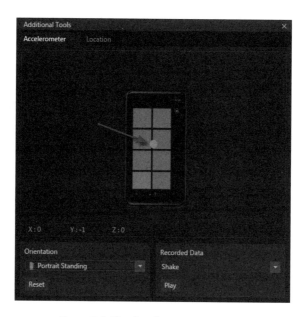

FIGURE 7.3 The Accelerometer window

can run. By default, the only recorded data that is included is Shake, which emulates someone shaking the phone for a few seconds.

Creating Recorded Data

You can add your recorded data to the emulator by creating a simple XML file and dropping it into the emulator's accelerometer sensor data directory. The directory is:

```
%PROGFILES%\Microsoft XDE\1.0\sensordata\acc
```

The XML files are labeled with the name you want to appear in the drop-down list (e.g., "Shake" with no extension for the Shake data file). The format of this XML file is shown below:

```
<?xml version="1.0" encoding="utf-8"?>
<WindowsPhoneEmulator
xmlns="http://schemas.microsoft.com/WindowsPhoneEmulator/2009/08/
SensorData">
  <SensorData>
    <Header version="1"/>
    <AccData offset="1000" x="1" y="0" z="0" />
    <AccData offset="2000" x="-1" y="0" z="0" />
    <AccData offset="3000" x="0" y="-1" z="0" />
```

```
    </SensorData>
</WindowsPhoneEmulator>
```

The important part of the format is the `SensorData` section. After the `Header` element is a list of `AccData` elements that contain the axis values as the number of milliseconds after the start of the recording to change the value to. This example moves the axis values once per second, but for something more complex you may need to change the values much more often than that. The Shake offsets change the values every few milliseconds to mimic the Shake action.

Using Sound

Like haptic feedback, using sound effectively can help you create a great application. As with any technique, though, some finesse is required to ensure optimal use. On Windows Phone, you have two options for playing sound: You can use the `MediaElement` or use XNA. In addition, the phone allows you to record sound. This section will cover all three of these aspects of using sound in your application.

Playing Sounds with `MediaElement`

Because sound is just another type of media, the easiest way for most people to play sound is to just use the `MediaElement`:

```
<Grid x:Name="ContentPanel">
  <MediaElement Name="soundElement"
                Source="/Ringtone.wma"
                AutoPlay="False" />
</Grid>
```

For playing sounds, the `MediaElement` is just an invisible part of the XAML and has no user interface. To use the `MediaElement`, you typically specify a source file (as a URL to the sound). By setting `AutoPlay` to `false` you can play the sound with code:

```
private void playButton_Click(object sender, RoutedEventArgs e)
{
    soundElement.Play();
}
```

While using a `MediaElement` is a straightforward way to play a sound, it has some drawbacks.

- You must tie the playing of the sound directly to the user interface (the `MediaElement` must exist in the XAML).
- Only one `MediaElement` can be playing a sound at a time. So playing different sounds means swapping the `Source` property as you use it.

When using the `MediaElement` you can use one of a number of formats, including the following:

- PCM WAV file (.wav)
- Microsoft Windows Media Audio (.wma)
- ISO MPEG-1 Layer III (.mp3)
- Unprotected ISO Advanced Audio Coding (.aac)

By using the `MediaElement` to play your audio files, you have full media control (e.g., volume, stereo control, play/pause/stop, position). In most cases, if you want to play longer sounds like songs or albums, you should probably use the built-in media player functionality, discussed next.

Using XNA Libraries

At the center of a standard XNA library is a game loop that runs for the length of the program and handles the task of updating the UI and accepting input. It is not event-based, like Silverlight. The game loop's job is to give up control so that each frame of the game can be shown (i.e., it updates the screen). This doesn't happen in Silverlight, so we must use the `FrameworkDispatcher` class's `Update` method to accomplish this for us. In some cases we can just call it once when we need to accomplish something with the XNA libraries (such as playing a sound effect); other times we need to mimic the updating. We can do this by using a `DispatcherTimer` to update the game loop periodically:

```
DispatcherTimer _updateTimer = new DispatcherTimer();

// Constructor
public MainPage()
{
  InitializeComponent();

  // Set up timer to call the XNA Dispatcher (e.g. Game Loop)
  _updateTimer.Interval = TimeSpan.FromMilliseconds(50);
  _updateTimer.Tick += (s, e) =>
    {
      FrameworkDispatcher.Update();
    };
}
```

When we do this, we are periodically ceding to the XNA framework to do the work it normally does during the game loop.

Playing Sounds with XNA

Because you're using Silverlight to build your applications for the phone, you also can use some of the XNA libraries in your application. The key class is the `SoundEffect` class, which lets you play a short, fire-and-forget sound. This class only supports PCM WAV (.wav) files, but it will give you a high-performance sound effect when you need it (it is "high-performance" in that it does not require nearly the amount of system resources to play a sound as do formats that require decompression [e.g., mp3 and wma]). In addition, playing sounds with XNA will enable you to play multiple sounds at the same time.

To use the `SoundEffect` class you'll need a reference to the `Micro-soft.Xna.Framework.dll` assembly and you'll have to import the `Microsoft.Xna.Framework.Audio` namespace. Once you do that you can create a sound effect by calling the `SoundEffect.FromStream` method, like so:

```
using Microsoft.Xna.Framework.Audio;

...

private void xnaPlayButton_Click(object sender, RoutedEventArgs e)
{
  // Get the sound from the XAP file
  var info = App.GetResourceStream(
    new Uri("alert.wav", UriKind.Relative));
```

```
// Load the SoundEffect
SoundEffect effect = SoundEffect.FromStream(info.Stream);

// Tell the XNA Libraries to continue to run
FrameworkDispatcher.Update();

// Play the Sound
effect.Play();
}
```

After you load the file from the XAP (using the `GetResourceStream` method of the `Application` class), you can create the sound effect directly from the stream of the .wav file. Before you can play the effect, you must tell XNA to let updates happen to the rest of the application using the `FrameworkDispatcher.Update` method. Finally, you can play the effect. Playing the effect is not a blocking call (e.g., `Play` returns immediately).

Adjusting Playback

You can also use the `SoundEffect` class to control the volume, pitch, and looping of a sound effect. To do this, you need to create an instance of the sound effect you've loaded, like so:

```
private void loopPitchButton_Click(object sender, RoutedEventArgs e)
{
  // Get the sound from the XAP file
  var info =
    App.GetResourceStream(new Uri("alert.wav", UriKind.Relative));

  // Load the SoundEffect
  SoundEffect effect = SoundEffect.FromStream(info.Stream);

  // Get an instance so we can affect the sound effect
  SoundEffectInstance instance = effect.CreateInstance();

  // Change the pitch and volume
  instance.Pitch = -.35f; // Slow it down
  instance.Volume = .95f; // Make it a little quiet

  // Loop the sound
  instance.IsLooped = true;

  // Tell the XNA Libraries to continue to run
  FrameworkDispatcher.Update();
```

```
// Play the Sound
instance.Play();
}
```

Once you create the instance, you can specify the pitch (zero is normal speed) and volume (1.0 is normal volume). You can also specify whether the sound is to be looped. After you've made these changes, you can play the sound with the changes (using the `Play` method). Note that you're playing the instance and not the sound effect, which is different from earlier examples.

Recording Sounds

Using the XNA libraries you can also record sound. To accomplish this you can use the `Microphone` class (in the `Microsoft.Xna.Framework .Audio` namespace), which provides access to the phone's microphone. The `Microphone` class has a static property (called `Default`) that returns the default microphone. This should be the only microphone that a phone has, so you should be able to reliably use this microphone. You will also need to store the results of the recording. The easiest way to do this is to use a `Stream` to store the information. In this case, a `MemoryStream` is perfect for just storing the recording in memory:

```
public partial class MainPage : PhoneApplicationPage
{
  MemoryStream _recording = null;
  DispatcherTimer _updateTimer = new DispatcherTimer();

  // Constructor
  public MainPage()
  {
    InitializeComponent();

    // Set up timer to call the XNA Dispatcher (e.g. Game Loop)
    _updateTimer.Interval = TimeSpan.FromMilliseconds(50);
    _updateTimer.Tick += (s, e) =>
      {
        FrameworkDispatcher.Update();
      };
  }
```

Note that I am using the `DispatcherTimer` to make sure we can use the XNA libraries (as detailed above). Next, you need to turn the microphone on and off:

```
private void recordButton_Click(object sender, RoutedEventArgs e)
{
  // Create a new Memory Stream for the data
  _recording = new MemoryStream();

  // Start the timer to create the 'game loop'
  _updateTimer.Start();

  // Start Recording
  Microphone.Default.Start();
}

private void stopButton_Click(object sender, RoutedEventArgs e)
{
  // Stop Recording
  Microphone.Default.Stop();

  // Stop the 'game loop'
  _updateTimer.Stop();
}
```

You can see these event handlers are starting and stopping not only the microphone but also the "game loop," which enables the recording APIs to work. When a new recording is started, a new `MemoryStream` is created to store the recording so that we get a new one for every recording.

The `Microphone` class has a `BufferReady` event that can accept the data from the microphone. As the buffer fills with the recording, the event will fire with a small amount of sound data that you need to store:

```
public partial class MainPage : PhoneApplicationPage
{
  MemoryStream _recording = new MemoryStream();
  DispatcherTimer _updateTimer = new DispatcherTimer();

  // Constructor
  public MainPage()
  {
    InitializeComponent();

    // Set up timer to call the XNA Dispatcher (e.g. Game Loop)
    _updateTimer.Interval = TimeSpan.FromMilliseconds(50);
    _updateTimer.Tick += (s, e) =>
      {
        FrameworkDispatcher.Update();
      };
```

```
// Wire up an event to get the data from the Microphone
Microphone.Default.BufferReady +=
  new EventHandler<EventArgs>(_mic_BufferReady);


}

void _mic_BufferReady(object sender, EventArgs e)
{
  // Grab the Mic
  var mic = Microphone.Default;

  // Determine the #/bites needed for our sample
  var bufferSize = mic.GetSampleSizeInBytes(mic.BufferDuration);

  // Create the buffer
  byte[] buffer = new byte[bufferSize];

  // Get the Data (and return the number of bytes recorded)
  var size = Microphone.Default.GetData(buffer);

  // Write the data to our MemoryStream
  _recording.Write(buffer, 0, size);
}
```

As you can see, the `BufferReady` event handler first retrieves the size of the buffer (by asking the `Microphone` class to get the sample size in bytes). Next it creates a new buffer of bytes to store the data. Then it calls the `Microphone` class to get the data (it takes the data, copies it into the buffer that is passed in, and returns the number of bytes that were used). Lastly, it uses the `MemoryStream` we created earlier and writes the new data into the stream. This event will be called enough times to store the data as it is being recorded.

Once the recording is complete, you can do whatever you want with the stream of sound. The `Microphone` class returns the data as a .wav file, so you can store it in isolated storage and use it later or just play it back using the `SoundEffect` API mentioned earlier:

```
private void playBackButton_Click(object sender, RoutedEventArgs e)
{
  if (Microphone.Default.State == MicrophoneState.Stopped &&
      _recording != null)
  {
    // Load the SoundEffect
    SoundEffect effect = new SoundEffect(_recording.ToArray(),
      Microphone.Default.SampleRate,
      AudioChannels.Mono);
```

```
    // Tell the XNA Libraries to continue to run
    FrameworkDispatcher.Update();

    // Play the Sound
    effect.Play();
  }
}
```

You can see here that the `Microphone` class also has a `State` property that you can use to make sure you only use the data once the recording is over. Then, instead of loading the `SoundEffect` object from a stream (as shown earlier), this code creates a new `SoundEffect` passing in the contents of the stream, the sample rate (which the `Microphone` class contains), and the number of audio channels. Otherwise, this code is just like playing any other sound effect.

Contacts and Appointments

Even though we can get caught up in the mystique of our phones as being little computers, most users actually use their phones to contact people and to keep their appointments. I know this must come as a big surprise to most developers, but it's true!

So, as a developer, you may want to access the user's contacts and appointments in order to more tightly integrate your applications into the phone experience. To do this, the Windows Phone SDK supports entry points into the user's contacts and appointments. These APIs provide a way to search by commonly used search semantics, but also return results that can be further filtered using LINQ.

> ■ **Note**
>
> All access to the contacts and appointments is read-only. You cannot change, add, or delete any information in the list of contacts or appointments.

Contacts

To access the contacts on a phone, you will start with the `Contacts` class (in the `Microsoft.Phone.UserData` namespace). This class provides the ability to perform an asynchronous search for the contacts on the phone.

```
var contactCtx = new Contacts();

contactCtx.SearchCompleted += (s, a) =>
  {
    IEnumerable<Contact> contacts = a.Results;
    contactList.ItemsSource = results;
  };

contactCtx.SearchAsync(null, FilterKind.None, null);
```

The `SearchAsync` call takes three parameters: a filter string, a type of filter, and an object that is passed to the completed event (a state object). The first two parameters typically allow you to do basic filtering in the underlying data store and are specified as the most common filtering semantics. The `FilterKind` enumeration tells the search API how to treat the filter string. For example, to find a specific person you can search by name, like so:

```
contactCtx.SearchAsync("Chris Sells", FilterKind.DisplayName, null);
```

The `FilterKind` enumeration includes the members shown in Table 7.1.

TABLE 7.1 `FilterKind` **Enumeration**

Value	Description
None	No filter is going to be used. The first parameter of `SearchAsync` is ignored.
DisplayName	Searches for contact based on the display name of the user. This type of filter supports substring matches (e.g., "Sells" matches "Chris Sells").
EmailAddress	Searches for contact based on email address. Only supports full email address in the filter.
PhoneNumber	Searches for contact based on phone number. Must be complete phone number, but format is not important (e.g., "2065550003" is the same as "(206) 555 0003").
PinnedToStart	Searches for contact that is pinned to the Start screen. The first parameter of `SearchAsync` is ignored.

To use the `FilterKind` enumeration, you can specify the filter and the `FilterKind` enumeration in the `SearchAsync` method:

```
contactCtx.SearchAsync("chris@example.com",
                       FilterKind.EmailAddress,
                       null);
```

The result of a search is an `IEnumerable` collection of `Contact` objects. The `Contacts` class supports all the information about a particular contact. Since the data structure can be verbose (e.g., there can be multiple phone numbers, addresses, company names, etc.), the class supports many collections for the different items in the data store. For example, you can get the simple display name as a property, but to get the phone number of a contact you have to use LINQ against the collection:

```
IEnumerable<Contact> contacts = a.Results;

var contact = contacts.First();

string displayName = contact.DisplayName;
string phoneNumber = contact.PhoneNumbers.First().PhoneNumber;
```

This means that to meaningfully bind to the `Contact` you will likely need to create a class that represents the data you want to show:

```
public class ContactInfo
{
  public string Name { get; set; }
  public ContactPhoneNumber PhoneNumber { get; set; }
  public DateTime? Birthday { get; set; }
  public ContactCompanyInformation Company { get; set; }
  public ImageSource Picture { get; set; }
  public ContactEmailAddress EmailAddress { get; set; }
}
```

Although you can have the bindable class only include basic types, you can also use the built-in data structures (e.g., `ContactPhoneNumber`) since they do support binding. Luckily, building a collection of these bindable objects is easy using LINQ projections:

```
IEnumerable<Contact> contacts = a.Results;

var qry = from c in contacts
          select new ContactInfo()
```

```
                        {
                            Name = c.DisplayName,
                            EmailAddress = c.EmailAddresses.FirstOrDefault(),
                            PhoneNumber = c.PhoneNumbers.FirstOrDefault(),
                            Company = c.Companies.FirstOrDefault(),
                        };

    var results = qry.ToList();
```

By using this technique of projecting into a new list of the contact information, you can easily build a `ListBox` using data binding:

```xml
<ListBox x:Name="contactList">
  <ListBox.ItemTemplate>
    <DataTemplate>
      <StackPanel Width="470">
        <TextBlock Text="{Binding Name}" />
        <TextBlock Text="{Binding EmailAddress.EmailAddress}" />
        <TextBlock Text="{Binding Company.CompanyName}" />
        <TextBlock Text="{Binding PhoneNumber.PhoneNumber}" />
      </StackPanel>
    </DataTemplate>
  </ListBox.ItemTemplate>
</ListBox>
```

You can see in the data binding that you can navigate down the data structure to the object you want to bind to. And since they are data bindings, if the user does not have one of these objects, the binding silently fails (which is one of the advantages of doing this in a binding).

Of course, since you can use LINQ to query against the resultant collection, you can use it if the `FilterKind` enumeration does not give you the granularity you need to find contacts. While this works well, be aware that loading all contacts in memory may consume a lot of memory. For example, if you want to find all the contacts who have a phone number in the 206 area code (as well as sort them by display name) you can just use LINQ:

```
    var qry = from c in contacts
              where c.PhoneNumbers
                    .Any(phone => phone.PhoneNumber.Contains("(206)"))
              orderby c.DisplayName
              select new ContactInfo()
                  {
                      Name = c.DisplayName,
                      EmailAddress = c.EmailAddresses.FirstOrDefault(),
```

```
                PhoneNumber = c.PhoneNumbers.FirstOrDefault(),
                Company = c.Companies.FirstOrDefault(),
            };
```

In addition to the basic contact information, you can also ascertain which accounts the user contacts. The `Contacts` class exposes a list of the user's accounts. Each account will tell you the kind of account (e.g., Facebook, Windows Live, Outlook, etc.) as well as the name of the account:

```
Account first = contactCtx.Accounts.FirstOrDefault();
if (first != null)
{
  string accountName = first.Name;
  if (first.Kind == StorageKind.Facebook)
  {
    MessageBox.Show("Your first account is Facebook");
  }
}
```

You cannot filter by account through the `SearchAsync` method, but you can filter any of the associated objects by account. For example, to filter contacts by account, simply use the following:

```
Account first = contactCtx.Accounts.FirstOrDefault();

var qry = from c in contacts
          where c.Accounts.Contains(first)
          select new ContactInfo()
              {
                Name = c.DisplayName,
                EmailAddress = c.EmailAddresses.FirstOrDefault(),
                PhoneNumber = c.PhoneNumbers.FirstOrDefault(),
                Company = c.Companies.FirstOrDefault(),
              };
```

Most of the objects related to contacts also have an account associated with them. For example, if you have a contact you can find out which email address is from Facebook, like so:

```
Contact theContact = contacts.First();

var addr = theContact.EmailAddresses
  .Where(email => email.Accounts
                      .Any(a => a.Kind == StorageKind.Facebook))
  .FirstOrDefault();
```

Lastly, each contact may have a picture associated with it. This picture is not immediately returned with the contact (as that could occupy a large chunk of memory). To access the picture, the `Contacts` class has a `GetPicture` method that returns a `Stream` object containing the image. While this is useful, you'll probably need to wrap it with an `ImageSource` object to use it in data binding. One approach is to do this during the projection into the bindable class, like so:

```
var qry = from c in contacts
          select new ContactInfo()
                 {
                     Name = c.DisplayName,
                     EmailAddress = c.EmailAddresses.FirstOrDefault(),
                     PhoneNumber = c.PhoneNumbers.FirstOrDefault(),
                     Company = c.Companies.FirstOrDefault(),
                     Picture = CreateImageSource(c.GetPicture())
                 };
```

The implementation of the `CreateImageSource` method that does the wrapping looks like this:

```
ImageSource CreateImageSource(Stream stream)
{
  if (stream == null) return null;
  var src = new BitmapImage();
  src.SetSource(stream);
  return src;
}
```

Because data binding can't turn a `Stream` into an image automatically, either you need to do that when you create the object or perhaps you can do it with a value converter (see Chapter 3, XAML Overview, for more information on value converters).

■ **Looking Forward**

Even though the contacts API is read-only, you can add new contacts via the `SaveContactTask` as detailed later in this chapter.

Appointments

Working with appointments follows the same pattern. It all starts with the `Appointments` class (like the `Contacts` class started with `Contacts`):

```
var appointmentCtx = new Appointments();

appointmentCtx.SearchCompleted += (s,a) =>
{
  IEnumerable<Appointment> appts = a.Results;

  // Bind to the UI
  apptList.ItemsSource = appts;
};

appointmentCtx.SearchAsync(DateTime.MinValue,
                           DateTime.Today,
                           null);
```

After creating the `Appointments` class, you can handle the `Search-Completed` event and call the `SearchAsync` method to perform a search. The `SearchAsync` method allows you to specify a date range for the search (the last parameter is an optional piece of state to send to the completed event). The data returned by the search is an enumerable list of `Appointment` objects.

Unlike the `Contacts` class, the `Appointments` class supports binding since the nature of the data is fairly simple. For example, by using data binding, a `ListBox` to show appointments could look as simple as this:

```
<ListBox x:Name="apptList">
  <ListBox.ItemTemplate>
    <DataTemplate>
      <StackPanel Width="470">
        <TextBlock Text="{Binding StartTime, StringFormat=d}" />
        <TextBlock Text="{Binding IsPrivate}" />
        <TextBlock Text="{Binding Location}" />
        <TextBlock Text="{Binding Subject}" />
        <TextBlock Text="{Binding Status}" />
      </StackPanel>
    </DataTemplate>
  </ListBox.ItemTemplate>
</ListBox>
```

The `SearchAsync` method also allows you to specify an account to search as well as the maximum number of results. So you can pick an account and then search like so:

```
var outlookAccount = appointmentCtx.Accounts
                        .Where(a => a.Kind == StorageKind.Outlook)
                        .FirstOrDefault();
```

```
appointmentCtx.SearchAsync(DateTime.MinValue,  // Start
                           DateTime.Today,      // End
                           25,                  // Max Results
                           outlookAccount,      // Account
                           null);
```

By using these APIs, you can access the contacts and appointments for the phone's user.

Alarms and Reminders

When you're creating your application, you may need the ability to alert the user about specific notifications. Windows Phone 7.5 supports this by allowing you to create two types of notifications: alarms and reminders.

Alarms are notifications that show a message and allow the user to tap the alarm to start your application. Additionally, the user can snooze the alarm or dismiss it. With an alarm you can also specify a custom sound located in your application to use as the alarm sound. You can see an alarm in action in Figure 7.4.

FIGURE 7.4 An alarm

Reminders are similar to alarms but they have additional functionality.

- You can specify the title of a reminder (which is always "Alarm" for alarms).
- The user can specify how long to snooze a reminder.
- You can decide to send contextual information when the user taps a reminder (whereas the alarm can simply launch the application).
- You cannot specify a specialized sound for reminders.

Figure 7.5 shows a reminder. If alarms or reminders are not closed when additional notifications appear, they will be stacked with an identifier that mentions how many notifications need to be handled, as shown in Figure 7.6.

Alarms and reminders are both tied to your application. They are notifications for your application. If you want to allow users to disable/enable these notifications, you must supply that functionality in your application. There is no operating system level management UI for these notifications.

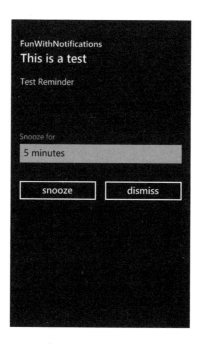

Figure 7.5 A reminder

FIGURE 7.6 Stacked notifications

> ### ■ Tip
>
> The precision of alarm and reminder times is to the minute. If you create notifications that need higher precision, the service that handles alarms and reminders will not show the notification more than once per minute. Be prepared for the alarms that are within a minute of a notification to be stacked.

Now that you've seen how notifications look, let's discuss how you actually create your own notifications.

Creating an Alarm

To create a new alarm, you can simply create a new `Alarm` object and instantiate all the relevant properties, like so:

```
// Create Alarm - name must be unique per app
var alarmName = Guid.NewGuid().ToString();

var alarm = new Alarm(alarmName)
{
```

```
  // When the Alarm should sound
  BeginTime = DateTime.Parse("2011-12-25T06:30"),

  // The Message in the Alarm
  Content = "Wake Up for Christmas!",

  // What sound to play for the alarm
  Sound = new Uri("santa.wav", UriKind.Relative)
};

// Add the Alarm to the Phone
ScheduledActionService.Add(alarm);
```

Each alarm must have a name that is unique to your application (i.e., names do not need to be globally unique). So it is common to just use a GUID to name your alarm. This name is not visible to the user. Once you construct the alarm using the unique name, you would typically specify several key properties, including the following.

- **BeginTime:** This is the time the alarm will alert the user.
- **Content:** This is the text to be shown in the alarm beneath the "Alarm" title.
- **Sound:** This is a URI to the sound in your .xap file that should be used to play when the alarm occurs.

Finally, you will use the `ScheduledActionService` to add the alarm to the phone. This service is the starting point for both alarms and reminders (as they are both considered a `ScheduledAction`). You can simply add your alarm to the phone by using the service class as shown in the example.

You can also create alarms to have recurrence. To do this you should set both the `RecurrenceType` and the `ExpirationTime`. The `Recurrence-Type` is an enumeration that allows you to specify whether the alarm is hourly, daily, weekly, and so forth. The `ExpirationTime` specifies how long to repeat the recurrence.

```
// ...
// Set Recurrence
alarm.RecurrenceType = RecurrenceInterval.Yearly;
alarm.ExpirationTime = DateTime.Today.AddYears(10);

// ...
```

The `Alarm` class has a `Title` property that looks like you can set the title of your alarm, but this is not supported (i.e., it throws a `NotSupported-Exception`). The title of all alarms is "Alarm".

Creating a Reminder

Since alarms and reminders both derive from `ScheduledAction`, you can correctly assume that creating a reminder is similar to creating an alarm:

```
// Create Reminder - name must be unique per app
var reminderName = Guid.NewGuid().ToString();

var reminder = new Reminder(reminderName)
{
  // Reminder Time
  BeginTime = DateTime.Today.AddHours(6),

  // Title for the Reminder
  Title = "Wake Up Reminder",

  // The description below the Title
  Content = "You should get up now...",

  // The page to navigate to (and any other context)
  // when the user taps the reminder
  NavigationUri =
    new Uri("/MainPage.xaml?wake=true", UriKind.Relative),
};

// Add the reminder
ScheduledActionService.Add(reminder);
```

However, creating a reminder is different from creating an alarm, in several ways. First, the `Title` property is supported, so you can specify what the title is on your reminder. Unfortunately, unlike the alarm, the `Sound` property is missing, so reminders can only have the sound that the system defines as the reminder sound (though the user can set that in the phone's Options section).

The biggest difference between an alarm and a reminder is that you can specify the `NavigationUri` to specify the deep linking of the reminder to a specific page (with context). The `NavigationUri` allows you to specify a relative URL of the page to show in your applications as well as an optional query string to include context to the reminder. If the user taps on the reminder, it will open your app and navigate to the URI you've specified.

Accessing Existing Notifications

Once you've added notifications to the phone, you may want to support managing those notifications. The `ScheduledActionService` class has several members that can help you manage your notifications.

- **Add:** This adds a new notification to the phone.
- **Find:** This locates a notification by name.
- **GetActions<T>:** This returns a read-only collection of notifications based on the type (alarm or reminder).
- **Remove:** This removes a specifically named notification from the phone.
- **Replace:** This replaces a named notification with a new notification (of the same name).

For example, to get a list of reminders and bind them to a list box, you can simply use the `GetActions` method:

```
// Returns an IEnumerable<Reminder> object
var reminders = ScheduledActionService.GetActions<Reminder>();

// A ListBox named 'theBox'
theBox.ItemsSource = reminders;
```

The base class for the notifications (`ScheduledAction`) supports a read-only property called `IsScheduled`, which is used to tell you if the alarm or reminder is still scheduled to show a notification in the future. Although the class supports the `IsEnabled` property, it is always `true` for alarms and reminders.

> ■ **Looking Forward**
>
> You can also use the `ScheduledActionService` and `Scheduled-Action` classes to create scheduled agents. We will discuss this in Chapter 9, Multitasking.

Using Tasks

Windows Phone OS 7.1 has several of the features of Windows Phone, including the address book, Web browser, and camera. To give you access to these features, the Windows Phone SDK 7.1 supports a list of tasks that you can execute. These tasks are divided into two categories: launchers and choosers. A **launcher** simply takes the user to another part of the phone experience to accomplish some goal (e.g., launch a Web page, make a phone call). A **chooser** takes the user to a facility on the phone but returns to the app with some information (e.g., choose an email address, take a picture). Tables 7.2 and 7.3 show the launchers and choosers available for the phone.

All of these tasks are in the same namespace: `Microsoft.Phone.Tasks`.

TABLE 7.2 Launchers

Launcher	Description
BingMapsTask	Displays a map of the result of a search or a specific coordinate
BingMapsDirectionsTask	Displays directions based on a start and end location
EmailComposeTask	Launches a new email to be sent by the user
MarketplaceDetailTask	Launches the details page for a specific Marketplace item (app or music)
MarketplaceHubTask	Launches a specific hub of the Marketplace
MarketplaceReviewTask	Launches a specific apps review page (usually your application)
MarketplaceSearchTask	Launches the search page for the user to search through the Marketplace
MediaPlayerLauncher	Launches the media player (e.g., Zune player)
PhoneCallTask	Launches the phone call user interface to make a phone call
SearchTask	Launches the search UI on the phone
ShareLinkTask	Shares a link with a social media application

TABLE 7.2 Launchers (*continued*)

Launcher	Description
ShareStatusTask	Shares the user's current status with a social media application
SmsComposeTask	Launches a new SMS message to be sent by the user
WebBrowserTask	Launches the Web browser

TABLE 7.3 Choosers

Chooser	Description
AddressChooserTask	Allows the user to pick a physical address from the user's address book
CameraCaptureTask	Launches the camera to allow the user to take a camera picture and return that taken picture to the app
EmailAddressChooserTask	Allows the user to select an existing contact's email to be returned to the app
GameInviteTask	Invites users to the current gaming session
PhoneNumberChooserTask	Allows the user to pick a phone number to be returned to the app
PhotoChooserTask	Allows the user to pick an existing photo from the photo collection on the phone and be returned to the app
SaveContactTask	Saves a new contact into an address book
SaveEmailAddressTask	Allows the user to save a new email address to the phone and return the new address to the app
SavePhoneNumberTask	Allows the user to save a new phone number to the phone and return the new number to the app
SaveRingtoneTask	Allows the user to save a sound file to the phone for use as a ring tone

Launchers

Each launcher works in a consistent manner. They create an instance of the launcher, set optional properties that will help the launcher, and call a `Show` method to take the user to the phone's functionality. For example, the code for showing the Microsoft website in the Web browser is as simple as this:

```
void webBrowserTask_Click(object sender, RoutedEventArgs e)
{
  var task = new WebBrowserTask();
  task.Uri = new Uri("http://microsoft.com");
  task.Show();
}
```

Although the launchers do not use a common interface, they do follow the same convention, so they should be easy to work with. Let's look at each one.

BingMapsTask

When your application needs to show a map, the `BingMapsTask` class can accomplish that for you. This task will launch the phone's map application with the map information you're looking for. With this task you can either show a specific geo-coordinate location (e.g., latitude and longitude) or use a search term to seed the map shown. To use a search term, just create the task, set the `SearchTerm` property, and show the task, as shown here:

```
var task = new BingMapsTask();

// Set the location
task.SearchTerm = "Turner Field";

task.Show();
```

The `SearchTerm` property can also be an address, as shown here:

```
var task = new BingMapsTask();

// Set the location
task.SearchTerm = "755 Hank Aaron Dr SE, Atlanta, Georgia 30315";

task.Show();
```

You can also use the `Center` property to specify a location. The `Center` property takes a `GeoCoordinate` object (which is defined in the `System.Device.Location` namespace from the `System.Device` assembly):

```
var task = new BingMapsTask();

// Set the location
task.Center = new GeoCoordinate(33.734083, -84.387526);

task.Show();
```

You must specify either the `Center` or the `SearchTerm` property. To further customize the map, you can tell the task what zoom level you want using the `ZoomLevel` property:

```
var task = new BingMapsTask();

// Set the location
task.Center = new GeoCoordinate(33.734083, -84.387526);

// Set the Zoom level (1-19)
task.ZoomLevel = 19;

task.Show();
```

The `ZoomLevel` property accepts any number greater than zero, though typically you would specify a zoom between 1 and 19. The higher the number, the more zoom the map shows.

BingMapsDirectionsTask

When you want to help the user of your application get directions, the `BingMapsDirectionsTask` comes to the rescue. This task allows you to specify the starting and ending locations for directions, and then launches the map application to show the directions. To do this, you need to create an instance of the `BingMapsDirectionsTask` and set the `Start` and `End` properties:

```
var task = new BingMapsDirectionsTask();

// Starting Location (Label and GeoCoordinate)
task.Start = new LabeledMapLocation(
  "Turner Field",
  new GeoCoordinate(33.734083, -84.387526));
```

```
// Ending Location (Label and GeoCoordinate)
task.End = new LabeledMapLocation(
  "Cooperstown",
  new GeoCoordinate(42.700073, -74.922864));

task.Show();
```

The `Start` and `End` properties take an object called `LabeledMap-Location`. This object has you specify a label to be shown to the user as well as a `GeoCoordinate` object (which is defined in the `System.Device.Location` namespace from the `System.Device` assembly) that specifies the location. The `GeoCoordinate` object specifies the location (typically in latitude and longitude). Showing this task takes the user straight to the directions part of Bing Maps.

EmailComposeTask

The `EmailComposeTask` allows the user to compose and send an email without necessarily sharing those details with your app. For this task, you can set most of the information about an outgoing email, including information in the To, CC, Subject, and Body sections:

```
void composeEmailButton_Click(object sender, RoutedEventArgs e)
{
  var task = new EmailComposeTask();
  task.To = "shawn@hotmail.com";
  task.Cc = "youremail@hotmail.com";
  task.Subject = "Your Email";
  task.Body = "This is an email";
  task.Show();
}
```

All of these properties are completely optional. If the user just wants to launch an email you can create it and call `Show`. The `Body` property is always plain text. There is no support for creating a body with HTML in this release.

MarketplaceDetailTask

This task's job is to launch the Marketplace app to show the details of a specific application. While the task does allow you to show specific items using the `ContentIdentifier` and `ContentType` properties, it normally launches without specifying these properties, and will launch the details page for your own application:

```
void marketplaceDetailButton_Click(object sender, RoutedEventArgs e)
{
  // Show our Detail Page
  var task = new MarketplaceDetailTask();
  task.Show();
}
```

If you have a specific application content identifier (the GUID that represents the application[1]) you can specify it:

```
void marketplaceDetailButton_Click(object sender, RoutedEventArgs e)
{
  // Show our Detail Page
  var task = new MarketplaceDetailTask();
  task.ContentIdentifier = "a518bd6c-280e-e011-9264-00237de2db9e";
  task.ContentType = MarketplaceContentType.Applications;
  task.Show();
}
```

MarketplaceHubTask

The purpose of the `MarketplaceHubTask` is to go to the Marketplace application on the phone to allow the user to browse different items in the Marketplace. When using this task, you need to specify whether the user needs to navigate to music or to applications:

```
void marketplaceHubButton_Click(object sender, RoutedEventArgs e)
{
  // Show the Marketplace Hub
  var task = new MarketplaceHubTask();
  task.ContentType = MarketplaceContentType.Applications; // Or Music
  task.Show();
}
```

Note that this task does not work in the emulator as the Marketplace isn't intended to work in emulation mode.

---

1. The GUID for your application is in your WMAppManifest.xml file as the `ProductID` under the `App` element.

`MarketplaceReviewTask`

The `MarketplaceReviewTask` is used to take the user to the current application's review page to allow him to review the application. There are no properties to set for this particular task, as shown below:

```
void marketplaceReviewButton_Click(object sender, RoutedEventArgs e)
{
  // Show the Application Review Page for this application
  var task = new MarketplaceReviewTask();
  task.Show();
}
```

Because your application is not technically in the Marketplace during development, this task will take you to a review page; then it will complain that it cannot find the application. This is enough to test this task. Once your application has been deployed it will take the user to the correct review page.

`MarketplaceSearchTask`

The `MarketplaceSearchTask` allows you to send the user to the Search page of the Marketplace app specifying what types of search to perform as well as any keywords with which to prepopulate the search, as shown below:

```
void marketplaceSearchButton_Click(object sender, RoutedEventArgs e)
{
  // Show the Marketplace Search Page
  var task = new MarketplaceSearchTask();
  task.ContentType = MarketplaceContentType.Applications; // Or Music
  task.SearchTerms = "Shooting Games";
  task.Show();
}
```

`MediaPlayerLauncher`

The purpose of the `MediaPlayerLauncher` is to play media (audio and/or video) that comes from your application (instead of media that is in the media library). The media must be either in the application's install directory (e.g., in the .xap file) or in isolated storage. To specify the location, you need to supply the task's `Location` property, which takes an enumeration (`MediaLocationType`) with which you can specify either `Install` (which indicates the file was in the .xap file) or `Data` (which indicates the media file is in isolated storage). In addition, you have to specify the `Media` property, which takes a relative `Uri` that points to the media either in the .xap file or in isolated storage, like so:

```
void mediaPlayerButton_Click(object sender, RoutedEventArgs e)
{
  // Launch the Media Player
  var task = new MediaPlayerLauncher();
  task.Location = MediaLocationType.Install; // Or Data
  task.Media = new Uri("bear.wmv", UriKind.Relative);
  task.Controls = MediaPlaybackControls.All;
  task.Show();
}
```

In addition, you can specify which controls are shown in the media player. When playing with the `MediaPlayerTask`, the media player can show three different controls, as highlighted in Figure 7.7. The control on the left is the Rewind control, the one on the right is the Fast-Forward control, and the one in the center is for toggling between Pause and Play.

The `MediaPlaybackControls` includes a set of enumeration flags, so you can pick which controls are used. Table 7.4 shows the `MediaPlaybackControls` enumeration.

FIGURE 7.7 **Media player controls**

TABLE 7.4 `MediaPlaybackControls` **Enumeration**

Value	Description
All	Shows all the controls including Pause, Stop, Skip, Rewind, and Fast-Forward
None	Shows the media without any controls

continues

TABLE 7.4 `MediaPlaybackControls` **Enumeration (***continued***)**

Value	Description
Pause	Flag value for the Pause control
Stop	Flag value for the Stop control (not applicable to the `MediaPlayerTask`)
Skip	Flag value for the Skip control (not applicable to the `MediaPlayerTask`)
Rewind	Flag value for the Rewind control
FastForward	Flag value for the Fast-Forward control

Typically, you would show all controls. However, you can decide to use specific controls by mixing the flags, like so:

```
void mediaPlayerButton_Click(object sender, RoutedEventArgs e)
{
  // Launch the Media Player
  var task = new MediaPlayerLauncher();
  task.Location = MediaLocationType.Install; // Or Data
  task.Media = new Uri("bear.wmv", UriKind.Relative);
  task.Controls = MediaPlaybackControls.Pause |
              MediaPlaybackControls.Rewind;
  task.Show();
}
```

PhoneCallTask

As its name suggests, the `PhoneCallTask` allows you to perform a phone call on the phone. All you need to do is specify the phone number and name to display to the user, like so:

```
void phoneCallButton_Click(object sender, RoutedEventArgs e)
{
  // Make a phone call
  var task = new PhoneCallTask();
  task.DisplayName = "Lottery Headquarters";
  task.PhoneNumber = "(404) 555 1212";
  task.Show();
}
```

FIGURE 7.8 `PhoneCallTask` **confirmation**

When this task is launched, the user will be asked for permission to make the phone call, as shown in Figure 7.8.

SearchTask

The `SearchTask` class is used to direct the user to the phone's built-in search application (the one that launches when you press the Search button on the phone). You can specify a search query to be prefilled when the search is launched, as shown below:

```
void searchButton_Click(object sender, RoutedEventArgs e)
{
  // Launch the Search App
  var task = new SearchTask();
  task.SearchQuery = "XBox Games";
  task.Show();
}
```

ShareLinkTask

If you want your application to be able to share a link on social networking sites for which the user has registered the phone (e.g., Facebook, Twitter), this is the task for you. It allows you to specify the link to share, the title of the link (i.e., what to display in the hyperlink), and a message to include with the link:

```
private void ShareLink_Click(object sender, RoutedEventArgs e)
{
  // Share a link
  var task = new ShareLinkTask();
  task.LinkUri = new Uri("http://wildermuth.com");
  task.Title = "Shawn's Blog";
  task.Message = "I can't believe his head is turning!";

  task.Show();
}
```

The user must have a social networking account registered on the phone for this launcher to succeed. If he has more than one social networking account on the phone, it will display a list of what service to use (similar to when you create a new email with more than one account).

ShareStatusTask

Much like the `ShareLinkTask`, this launcher allows you to share your current status via social networking. This launcher has you specify the user's status to share:

```
private void ShareStatus_Click(object sender, RoutedEventArgs e)
{
  // Share a link
  var task = new ShareStatusTask();
  task.Status = "I'm writing a #wp7 book!";

  task.Show();
}
```

Once shown, this will take you to the share status UI for the phone.

SmsComposeTask

The `SmsComposeTask` allows you to get the user to send a text message (e.g., SMS message). You can specify both whom to send the text to and the body of the message. The `To` property specifies a semicolon-delimited list of recipients for the text message. The `Body` property allows you to specify the message to be sent via SMS.

```
void smsButton_Click(object sender, RoutedEventArgs e)
{
  // Send an SMS Message
  var task = new SmsComposeTask();
  task.To = "(404) 555-1212; (206) 555-1212";
  task.Body = "This is a text message!";
  task.Show();
}
```

> **■ Note**
>
> This task does not allow you to send MMS messages (SMS messages with attachments).

WebBrowserTask

The WebBrowserTask launches the built-in browser (Internet Explorer) with a specified URI, as shown below:

```
void webBrowserTask_Click(object sender, RoutedEventArgs e)
{
  var task = new WebBrowserTask();
  task.Uri = new Uri("http://microsoft.com");
  task.Show();
}
```

Choosers

Now that you've seen the launchers, let's look at the choosers. The choosers present a very similar development experience, but unlike the launchers, they expect to return to your application with some data. All of the choosers support an event (usually called Completed) that is fired when the chooser is complete (whether it succeeds or not). Because your application may be tombstoned when a chooser is launched, you have to wire up this event in such a way that it will be rewired when the application is untombstoned. Typically you would do this by having your task created at the page level and wired up during page initialization (e.g., in the constructor):

```
public partial class MainPage : PhoneApplicationPage
{
  CameraCaptureTask cameraCapture = new CameraCaptureTask();
```

```
// Constructor
public MainPage()
{
  InitializeComponent();

  // Wire up completed event so it survives tombstoning
  cameraCapture.Completed +=
    new EventHandler<PhotoResult>(cameraTask_Completed);
}
...
}
```

When you launch a chooser you can simply call `Show`, like you did with the launchers in the earlier examples:

```
void cameraButton_Click(object sender, RoutedEventArgs e)
{
  cameraCapture.Show();
}
```

In the event handlers of all the choosers you must check the `Error` property to make sure an exception wasn't thrown during the chooser operation. In addition, you should check the `TaskResult` property on each chooser to be sure that the chooser was not canceled. You can check this by checking the `TaskResult` against the `TaskResult` enumeration:

```
void cameraTask_Completed(object sender, PhotoResult e)
{
  if (e.Error == null && e.TaskResult == TaskResult.OK)
  {
    // ...
  }
}
```

The following subsections provide examples showing how to use each chooser.

AddressChooserTask

Sometimes you may need to allow the user to retrieve a physical address from the address book. That is what the `AddressChooserTask` is designed to do. You can wire up this task to let the user pick an address:

```
AddressChooserTask addressChooser = new AddressChooserTask();

private void addressButton_Click(object sender, RoutedEventArgs e)
{
```

```
    addressChooser.Show();
}

void addressChooser_Completed(object sender, AddressResult e)
{
  if (e.Error == null)
  {
    string addressName = e.DisplayName;
    string address = e.Address;
  }
}
```

When the task completes, it returns both the name of the address (the display name of the person/company in the address book) and the address as a string. The address will have line breaks embedded if it is multilined (as most are) so that they can be displayed directly in a `TextBlock` without additional formatting.

CameraCaptureTask

The `CameraCaptureTask` is intended to instruct your user to take a picture and return to you the raw results of the picture:

```
CameraCaptureTask cameraCapture = new CameraCaptureTask();

  void cameraButton_Click(object sender, RoutedEventArgs e)
{
  cameraCapture.Show();
}

void cameraTask_Completed(object sender, PhotoResult e)
{
  if (e.Error == null && e.TaskResult == TaskResult.OK)
  {
    // Create a BitmapImage from the photo
    BitmapImage bitmap = new BitmapImage();
    bitmap.SetSource(e.ChosenPhoto);

    // Paint the background with the bitmap
    ImageBrush brush = new ImageBrush();
    brush.ImageSource = bitmap;
    LayoutRoot.Background = brush;
  }
}
```

In the event handler that is called after the user takes a picture you are handed the photo (as the `PhotoResult`'s `ChosenPhoto` property) as a

`Stream` object. You can then use the results of the photo in any way you want (e.g., store it in isolated storage, upload it to a server, or show it in the UI).

◾ Looking Forward

For access to the camera in real time (instead of requesting a photo from the photo-taking application) you can use the camera APIs detailed later in this chapter.

EmailAddressChooserTask

The purpose of this chooser is to allow the user to pick an email address from his list of contacts on the phone. When you use this chooser, it only shows contacts with email addresses to pick from, and if a contact has more than one email address, it will let the user pick which one to use. To use the `EmailAddressChooserTask`, you simply handle the event and show the task:

```
void chooseEmailTask_Click(object sender, RoutedEventArgs e)
{
  // Pick an email address
  task.Show();
}

void task_Completed(object sender, EmailResult e)
{
  // Ensure no error and that an email was chosen
  if (e.Error == null && e.TaskResult == TaskResult.OK)
  {
    MessageBox.Show(e.Email);
  }
}
```

In the event handler that is called once a user picks an email, you can get at the email address via the `Email` property of the `EmailResult` object.

PhoneNumberChooserTask

Like the email chooser, this task is designed to get the user to give you a phone number from his contacts. It works in the same way as the email chooser:

```
void choosePhoneTask_Click(object sender, RoutedEventArgs e)
{
```

```
    // Get Phone Number from Contacts
    task.Show();
}

void task_Completed(object sender, PhoneNumberResult e)
{
    // Ensure no error and that a phone number was chosen
    if (e.Error == null && e.TaskResult == TaskResult.OK)
    {
      MessageBox.Show(e.PhoneNumber);
    }
}
```

Once the user has selected a phone number, the event returns with the selected phone number.

PhotoChooserTask

This chooser is used to let the user get a photo and return it to your application. At first glance it works just like the `CameraCaptureTask`, but instead it lets the user pick a picture from the ones stored on his phone:

```
void choosePhotoTask_Click(object sender, RoutedEventArgs e)
{
  // Get a Photo
  task.Show();
}

void photoChooser_Completed(object sender, PhotoResult e)
{
  if (e.Error == null && e.TaskResult == TaskResult.OK)
  {
    // Create a BitmapImage from the photo
    BitmapImage bitmap = new BitmapImage();
    bitmap.SetSource(e.ChosenPhoto);

    // Paint the background with the bitmap
    ImageBrush brush = new ImageBrush();
    brush.ImageSource = bitmap;
    brush.Stretch = Stretch.None;
    LayoutRoot.Background = brush;
  }
}
```

You can allow the user to take a photo instead of picking a photo from the phone by setting the `ShowCamera` property on the task:

```
void choosePhotoTask_Click(object sender, RoutedEventArgs e)
{
  // Get a Photo
  task.ShowCamera = true;
  task.Show();
}
```

If you specify that the `ShowCamera` property is `true`, the `Application-Bar` in the photo chooser will have a camera button to let the user take a photo instead.

The other option that is available to the chooser is to specify a photo size that you need for your application:

```
void choosePhotoTask_Click(object sender, RoutedEventArgs e)
{
  // Get a Photo
  task.PixelHeight = 200;
  task.PixelWidth = 300;
  task.Show();
}
```

By specifying the `PixelHeight` and `PixelWidth` properties, you are telling the task to return the photo in that exact dimension. When the user picks a photo that does not match that size, he is presented with the ability to crop the photo to the selected size, as shown in Figure 7.9.

FIGURE 7.9 Allowing photo cropping

SaveContactTask

You can also allow the user to save new contacts to the phone, using the SaveContactTask. To do this, you specify the contact information in the SaveContactTask object like so:

```
private void saveContact_Click(object sender, RoutedEventArgs e)
{
  saveContact.Company = "Tailspin Toys";
  saveContact.FirstName = "Walter";
  saveContact.LastName = "Harp";
  saveContact.MobilePhone = "(206) 555-0142";
  saveContact.PersonalEmail = "wharp@tailspintoys.com";

  saveContact.Show();
}
```

The SaveContactTask has properties to provide access to a large amount of a contact's information, including multiple phone numbers, multiple addresses, and a complete name. Any contact the user adds here will show up in the contacts API discussed earlier in this chapter. Note that you can add a contact and read contacts with the contacts API, but you can't currently edit a contact. The user has to do that.

SaveEmailAddressTask

The SaveEmailAddressTask is used to save an email into a new contact or attach it to an existing contact. To save an address, simply supply the email to save and call Show:

```
void saveEmailTask_Click(object sender, RoutedEventArgs e)
{
  // Save a new Email on the Phone
  saveEmailTask.Email = "shawn@aol.com";
  saveEmailTask.Show();
}

void saveEmailTask_Completed(object sender, TaskEventArgs e)
{
  if (e.Error != null)
  {
    MessageBox.Show("Error Occurred");
  }
  else if (e.TaskResult == TaskResult.Cancel)
  {
    MessageBox.Show("User Cancelled Task");
  }
```

```
    else
    {
      MessageBox.Show("E-mail Saved");
    }
  }
```

Note that the task does not tell you what contact the email was attached to, but simply tells you whether the task succeeded or not.

SavePhoneNumberTask

This chooser works just like `SaveEmailAddressTask`; you simply supply the phone number and then call the `Show` method to save the phone number to a user's contact:

```
void savePhoneTaskButton_Click(object sender, RoutedEventArgs e)
{
  // Save a new Email on the Phone
  savePhoneTask.PhoneNumber = "(404) 555-1212";
  savePhoneTask.Show();
}

void savePhoneTask_Completed(object sender, TaskEventArgs e)
{
  if (e.Error != null)
  {
    MessageBox.Show("Error Occurred");
  }
  else if (e.TaskResult == TaskResult.Cancel)
  {
    MessageBox.Show("User Cancelled");
  }
  else
  {
    MessageBox.Show("Phone Number Saved");
  }
}
```

Again, the `TaskResult` will indicate whether the task succeeded, but does not share with your application the contact the phone number was attached to.

SaveRingtoneTask

If you're in the business of selling ring tones, your application can add ring tones to the phone. For a ring tone to be added to the phone, you must use the `SaveRingtoneTask`. A ring tone must conform to certain requirements.

- It must not be more than 39 seconds in length.
- It must be no more than 1MB in size.
- It can be in MP3 or WMA format only.
- It must not include DRM copyright protection.

In addition to these requirements, you must save the ring tone to isolated storage. We will discuss how to work with isolated storage in Chapter 8, Databases and Storage. Once you have the file saved in isolated storage, you can use the `SaveRingtoneTask` to add the ring tone to the phone. You need to specify a `DisplayName` (though the user can override it when the task launches), a URI to the ring tone in isolated storage, and whether the ring tone is sharable. Here is the relevant code:

```
void addRingtone_Click(object sender, EventArgs e)
{
  saveRingtoneTask.IsShareable = true;
  saveRingtoneTask.DisplayName = "Ahhh...";
  saveRingtoneTask.Source = new Uri("isostore:/ahhh.mp3");
  saveRingtoneTask.Show();
}

void task_Completed(object sender, TaskEventArgs e)
{
  if (e.Error == null && e.TaskResult == TaskResult.OK)
  {
    MessageBox.Show("RingTone Added!");
  }
}
```

The URI syntax used in this task specifies that it is an absolute URI that begins with "isostore:/" to specify that it is in your isolated storage. You can have structure to the location in isolated storage, so "isostore:/ringtones/sounds/ahhh.mp3" is perfectly acceptable. The return call to the `Completed` handler assures you that there wasn't an error and that the ring tone was successfully added.

Media and Picture Hubs

The phone can store music, pictures, and video. From the user's perspective these are stored in the Music+Videos hub and the Pictures hub. You can work with these hubs in several ways. You can access the music in the Music+Videos hub; you can access the pictures in the Pictures hub; and you can register your application to be included in both of these hubs.

Accessing Music

You can access the music on the phone directly using the `MediaLibrary` class in the `Microsoft.Xna.Framework` assembly. This class belongs to the `Microsoft.Xna.Framework.Media` namespace. To use the `Media-Library` class, you have to create an instance of the class. The default (i.e., empty) constructor creates an instance that contains a list of the media on the phone itself. This class supports `IDisposable`, so you must use care in calling `Dispose` when you're done with the class (usually via the `using` clause):

```
using (var library = new MediaLibrary())
{
  // use media library
}
```

Emulator Tip

The emulator has a set of sample media (music and pictures) that you can use through the `MediaLibrary` class, but they *only* appear when you're running under the debugger. Accessing the properties of the `MediaLibrary` class yields empty collections when not running under the debugger.

The music library consists of the following types of objects.

- **Artist:** This is the name of the performer of a song or album.
- **Album:** This is a collection of songs from one or more artists.
- **Genre:** This is a named category for songs and albums.
- **Song:** This is a single piece of music that can belong to a genre, album, and artist.
- **Playlist:** This is a user-defined list of songs.

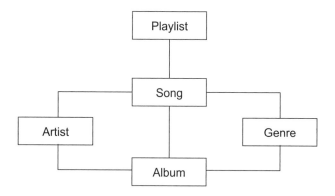

FIGURE 7.10 Music library objects

The `MediaLibrary` class exposes these types of objects into collections that can be navigated (as shown in Figure 7.10). While these are discrete collections, they each contain all the music in the collection.

■ MediaSource

The `MediaLibrary` class has a constructor that takes a `Media-Source`, so you may think that there is more than one `MediaSource` for the phone; however, this API exists for other XNA platforms. The phone has one and only one source (the local media library). *You should always use the empty constructor.*

On the `MediaLibrary` class, each of these objects is exposed as collections (that support `IEnumerable`), so simple iteration is straightforward:

```
using (var library = new MediaLibrary())
{
  // use media library
  foreach (var artist in library.Artists)
  {
    theList.Items.Add(artist.Name);
  }
}
```

Because these collections support `IEnumerable`, we can use LINQ against these as well:

```
using (var library = new MediaLibrary())
{
  var qry = from artist in library.Artists
            where artist.Name == "Shawn Twain"
            select artist;

  var shawn = qry.FirstOrDefault();

  if (shawn != null)
  {
    // Retrieve the songs for that artist.
    var hisSongs = shawn.Songs;
  }
}
```

As this example shows, once you retrieve the artist you're look-ing for you can use the songs for a particular artist, album, or genre. These collections are read-only, so you cannot add songs to the library programmatically.

Playing Music

Also built into XNA's media functionality is access to the media player on the phone, via the `MediaPlayer` class. This is a static class that gives you access to what is playing currently and allows you to queue up songs to be played. Since the class throws events, we need to update the `Framework-Dispatcher` (as we discussed earlier in this chapter):

```
public partial class MainPage : PhoneApplicationPage
{
  DispatcherTimer _xnaTimer = new DispatcherTimer();

  // Constructor
  public MainPage()
  {
    InitializeComponent();

    // Since we're using XNA's MediaPlayer, we need to
    // Use the FrameworkDispatcher to allow events
    _xnaTimer.Interval = TimeSpan.FromMilliseconds(50);
    _xnaTimer.Tick += (s, a) => FrameworkDispatcher.Update();
    _xnaTimer.Start();
  }
  ...
}
```

Once we do that, we can use the MediaPlayer class to play a song:

```
using (var library = new MediaLibrary())
{
  var qry = from artist in library.Artists
            where artist.Name == "Shawn Twain"
            select artist;

  var shawn = qry.FirstOrDefault();
  if (shawn != null)
  {
    // Retrieve the songs for that artist.
    var hisSongs = shawn.Songs;
    if (MediaPlayer.State != MediaState.Playing)
    {
      MediaPlayer.Play(hisSongs[0]);
    }
  }
}
```

You need to check the state of the MediaPlayer before you can play new songs, but as long as a song isn't currently playing, you can just play it by supplying the Song object to the media player. If you have songs you want to play that are not in the media library, you can create your own song objects using the FromUri static method:

```
var mySong = Song.FromUri("Bridges and Ghosts",
                      new Uri("song.mp3", UriKind.Relative));

if (MediaPlayer.State != MediaState.Playing)
{
  MediaPlayer.Play(mySong);
}
```

The FromUri method will take a relative URI that picks up songs in your .xap file or an absolute URI to add songs from anywhere on the Internet.

You can also supply a SongCollection to play the entire collection:

```
// Retrieve the songs for that artist.
var hisSongs = shawn.Songs;
if (MediaPlayer.State != MediaState.Playing)
{
  MediaPlayer.Play(hisSongs);
}
```

You are limited to playing only the SongCollections that already exist (e.g., MediaLibrary.Songs, Artist.Songs, Playlist.Songs, etc.). There is no facility for creating your own list of songs to integrate with the MediaPlayer class.

■ Playing Songs from Isolated Storage

If you want to play songs directly from isolated storage, you'll need to use the MediaElement, as the MediaPlayer class does not support this capability.

Accessing Pictures

In addition to accessing the music on the phone, the MediaLibrary class also exposes the pictures that are stored on the phone. The MediaLibrary class has three properties that give you access to the pictures on the phone.

- **Pictures:** This is a collection of all the pictures on the phone.
- **SavedPictures:** This is the special folder for saved pictures on the phone.
- **RootPictureAlbum:** This is the starting point for a hierarchical collection of pictures on the phone.

The Pictures and SavedPictures properties are simple IEnumerable collections of Picture objects. For example, to access the name of each picture you could iterate through the list of Picture objects like so:

```
using (var library = new MediaLibrary())
{
  foreach (Picture thePicture in library.Pictures)
  {
    theListBox.Items.Add(thePicture.Name);
  }
}
```

The Picture class exposes the name, date, height, and width of the picture as well as what album the picture belongs to. To get at the actual picture, you can call either GetThumbnailImage or GetImage to get a stream of the picture. GetThumbnailImage retrieves a much smaller version of the picture whereas GetImage retrieves the full-size picture. You can wrap the image you retrieve with a `BitmapImage` object from Silverlight's imaging system (in the `System.Windows.Imaging` namespace) to be able to show it as the source of an `Image` element, like so:

```
using (var library = new MediaLibrary())
{
  foreach (Picture thePicture in library.Pictures)
  {
    // Get the Picture Stream
    Stream imageStream = thePicture.GetImage();

    // Wrap it with a BitmapImage object
    var bitmap = new BitmapImage();
    bitmap.SetSource(imageStream);

    // Create an Image element and set the bitmap
    var image = new Image();
    image.Source = bitmap;

    // Add it to the ListBox to show it.
    theListBox.Items.Add(image);
  }
}
```

Although accessing individual pictures is useful, the `MediaLibrary` class also gives you access to the picture album structure. The `PictureAlbum` class contains a collection of that album's pictures (called `Pictures`) as well as a collection of the albums in that album (called `Albums`). The `MediaLibrary.RootPictureAlbum` is the top-level album and will contain a hierarchical collection of all albums and pictures. For example, you could iterate through all the albums using a recursive function, like so:

```
void MainPage_Loaded(object sender, RoutedEventArgs e)
{
  using (var library = new MediaLibrary())
  {
    AddAlbum(library.RootPictureAlbum, "");
  }
}

void AddAlbum(PictureAlbum theAlbum, string indention)
{
  // Show Album Name
  theListBox.Items.Add(string.Concat(indention,
                                     "Album: ",
                                     theAlbum.Name));
```

```
// List Albums in this Album
foreach (PictureAlbum subAlbum in theAlbum.Albums)
{
  AddAlbum(subAlbum, string.Concat(indention, "  "));
}

// List Pictures
foreach (Picture thePicture in theAlbum.Pictures)
{
  theListBox.Items.Add(string.Concat(indention,
                                     " - ",
                                     thePicture.Name));

}
}
```

Walking through the list of albums results in a list of the albums and their resultant pictures, as shown in Figure 7.11.

Storing Pictures

The `MediaLibrary` class also allows you to save pictures to the phone. The only limitation is that you can only save images directly to a special album called "Saved Pictures". You save the picture to this album using the `MediaLibrary.SavePicture` method. This method takes the name of the picture file as well as the contents of the picture (usually a stream or an array of bytes). For example, to capture a photo using the `CameraCaptureTask`

Figure 7.11 Displaying the albums and pictures

(explained in more detail earlier in this chapter), you could take the stream from the `Completed` event, like so:

```
public partial class MainPage : PhoneApplicationPage
{
  CameraCaptureTask takePicture = new CameraCaptureTask();

  // Constructor
  public MainPage()
  {
    InitializeComponent();

    MouseLeftButtonUp += new
      MouseButtonEventHandler(MainPage_MouseLeftButtonUp);


    // Handle the picture after it's been taken
    takePicture.Completed += new
      EventHandler<PhotoResult>(takePicture_Completed);
  }

  void MainPage_MouseLeftButtonUp(object sender,
                                  MouseButtonEventArgs e)
  {
    // Take the picture
    takePicture.Show();
  }

  void takePicture_Completed(object sender, PhotoResult e)
  {
    if (e.TaskResult == TaskResult.OK)
    {
      // Use the Media Library to save the picture
      using (MediaLibrary theLibrary = new MediaLibrary())
      {
        // The name supplied will be suffixed with .jpg
        theLibrary.SavePicture("My Camera Photo", e.ChosenPhoto);
      }
    }
  }
}
```

The `MediaLibrary` class's `SavePicture` method takes a name (that it suffixes with ".jpg") and the photo itself, which can be in the form of a stream (like this example shows) or a byte array. Once you save the picture it will show up in a new album called "Saved Pictures" and will be accessible from the `MediaLibrary` class's `SavedPictures` property.

Integrating into the Pictures Hub

Windows Phone OS 7.1 allows you to make an application that can register itself as an app in three places: the Pictures hub, picture viewer, and share picker. You can see the apps menu in the Pictures hub (under "apps") in Figure 7.12.

In order to integrate your application into these parts of the picture experience, you need to include a new section in the WMAppManifest.xml file. This section is called "Extensions":

```
<Deployment ...>
  <App ...>
    ...
    <Extensions>
      <!-- Integrate into pictures hub -->
      <Extension ExtensionName="Photos_Extra_Hub"
               ConsumerID="{5B04B775-356B-4AA0-AAF8-6491FFEA5632}"
               TaskID="_default" />
    </Extensions>
  </App>
</Deployment>
```

The `ExtensionName` and `ConsumerID` are both used to determine what type of integration your application will use (e.g., Pictures hub, picture viewer, picture sharing). The `TaskID` is used to determine how to launch your application. Using the `_default` indicates that your application should be launched normally and is typical.

FIGURE 7.12 The apps in the Pictures hub

As stated earlier, you can integrate into any or all of the three picture extensions. Each extension has its own entry in the Extensions section of the WMAppManifest.xml file:

```
<Deployment ...>
  <App ...>
    ...
    <Extensions>

      <!-- Integrate into pictures hub -->
      <Extension ExtensionName="Photos_Extra_Hub"
                 ConsumerID="{5B04B775-356B-4AA0-AAF8-6491FFEA5632}"
                 TaskID="_default" />

      <!-- Integrate into picture viewer -->
      <Extension ExtensionName="Photos_Extra_Viewer"
                 ConsumerID="{5B04B775-356B-4AA0-AAF8-6491FFEA5632}"
                 TaskID="_default" />

      <!-- Integrate into picture sharing -->
      <Extension ExtensionName="Photos_Extra_Share"
                 ConsumerID="{5B04B775-356B-4AA0-AAF8-6491FFEA5632}"
                 TaskID="_default" />

    </Extensions>
  </App>
</Deployment>
```

A picture extension application will be launched either from one of the picture integration locations (e.g., from the Pictures hub) or in the normal way (e.g., from the Start menu or app list). You will need to determine which way your application is being launched and show the chosen photo if launched via the Extras menu. This is typically handled by overriding the OnNavigatedTo method of your application's first page (e.g., MainPage.xaml) and using the NavigationContext property to test for the existence of a token that was passed to your application from the Pictures hub:

```
public partial class MainPage : PhoneApplicationPage
{
  // ...

  protected override void OnNavigatedTo(NavigationEventArgs e)
  {
    base.OnNavigatedTo(e);
```

```
// If token is in the query string,
// we launched here from Extras menu
if (NavigationContext.QueryString.ContainsKey("token"))
{
  var token = NavigationContext.QueryString["token"];

  // Use Media Library to open the image
  using (MediaLibrary library = new MediaLibrary())
  {
    Picture selectedPicture = library.GetPictureFromToken(token);

    // Use the Picture
    Stream picture = selectedPicture.GetImage();

    // Use the picture. Typically should show it on launch page
  }
}
// ...
}
```

Once you see the "token" in the query string, you can use that token to retrieve the picture from the MediaLibrary class by calling the Get-PictureFromToken method, as shown earlier.

Integrating into the Music+Videos Hub

If you are building an application that will be playing music or showing videos, you can integrate it into the Music+Videos hub. The main hub contains several areas into which you can integrate your application. The hub consists of four sections.

- **Zune:** This is the starting point for users to play music/videos/ podcasts that are synced onto their phones.
- **History:** This is a list of recently played items.
- **New:** This is a list of music and video items newly added to the phone.
- **Marquee:** This is a list of music and/or video applications on the phone.

Once you specify you are integrating with the phone, you should integrate with the History and New sections of this hub. By using the Media and Videos APIs, your application will automatically be listed in the Marquee section of the hub.

Debugging Music+Videos Hub Integration

Hub integration is enabled as your application is passed through the Marketplace certification process. To debug your application's integration you can add `HubType="1"` to the `App` element of the WMManifest.xml file, as shown below:

```
<?xml version="1.0" encoding="utf-8"?>
<Deployment
    xmlns="http://schemas.microsoft.com/windowsphone/2009/deployment"
    AppPlatformVersion="7.0">
  <App xmlns=""
      ProductID="{1f6ccd50-764b-4247-ba61-253a287ab640}"
      Title="FunWithMediaLibrary"
      RuntimeType="Silverlight"
      Version="1.0.0.0"
      Genre="apps.normal"
      Author="FunWithMediaLibrary author"
      Description="Sample description"
      Publisher="FunWithMediaLibrary"
      HubType="1">
```

Note that there is no way to run and test this behavior with the emulator. You must debug this on a physical device. It is recommended that you use the WPConnect.exe tool to connect the phone to the development computer as connecting with Zune will not allow you to have access to the Music+Videos hub (showing the "Syncing" message instead).

Integrating with the Now Playing Section

Whenever a user plays music or a video in your application, it is customary for you to tell the hub that a recently used piece of music or video was played. You accomplish that with the `MediaHistoryItem` class. This class contains information that will be used to show what media is now playing as well as information that is passed back to your application when it is launched through the hub. To use this class, you create an instance of this class with the information required and add it to the hub via the `MediaHistory` class:

```
// You will need an image for the Now Playing section of
// the Hub.
Stream songImage = GetImageForSong();

// Create an object that contains the history information
MediaHistoryItem item = new MediaHistoryItem();
item.Source = "";
item.ImageStream = songImage;
item.Title = "NowPlaying";
item.PlayerContext.Add("keyString", "mysong.mp3");
MediaHistory.Instance.NowPlaying = item;
```

When you create the `MediaHistoryItem`, the `Source` property must be an empty string. The `ImageStream` must be a JPEG image that is 358 pixels by 358 pixels in size. It should include your application's name and logo if any. For the Now Playing section of the Music+Videos hub, the `Title` must be "NowPlaying". The `PlayerContext` property is a name/value list of information that you will have access to when an item launches your application.

Integrating with the History Section

When any media is played, you should create a new `MediaHistoryItem` for each media item. When you add them to the History section they will show up in the History part of the Music+Videos hub. To do this, create a `MediaHistoryItem` like we did for the Now Playing section:

```
// Create an object that contains the history information
MediaHistoryItem item = new MediaHistoryItem();
item.Source = "";
item.ImageStream = songImage;
item.Title = "RecentPlay";
item.PlayerContext.Add("keyString", "mysong.mp3");
MediaHistory.Instance.WriteRecentPlay(item);
```

The differences between this and the Now Playing section are that the `Title` property must be `RecentPlay` and you must add it to the `Media-History` via the `WriteRecentPlay` method. Lastly, images for the History tab are different and must be 173 pixels by 173 pixels instead of the larger Now Playing size. You should not just resize the larger image as the text on that image would likely be unreadable at the smaller size.

Integrating with the New Section

When new media is added via your application, you will need to add `MediaHistoryItems` as you did for the other sections:

```
// Create an object that contains the new information
MediaHistoryItem item = new MediaHistoryItem();
item.Source = "";
item.ImageStream = songImage;
item.Title = "MediaHistoryNew";
item.PlayerContext.Add("keyString", "mysong.mp3");
MediaHistory.Instance.WriteAcquiredItem(item);
```

This works identically to adding history items except that the `Title` must be `"MediaHistoryNew"` and adding the item to the New section requires that you call the `WriteAcquiredItem` method. The size of the image is 173 pixels by 173 pixels, just like the History section example.

Handling Launching from the Hub

When an item is launched from the hub, your main page is launched with a query string parameter that matches the `PlayerContext` item you added above. What is in the `keyString` is entirely application-specific. What you put in the `PlayerContext`'s `keyString` value will be passed back to your application. Usually you would handle this in the `OnNavigatedTo` overridable method of your main page, like so:

```
protected override void OnNavigatedTo(NavigationEventArgs e)
{
  base.OnNavigatedTo(e);

  // If "keyString" is in the query string, we were launched
  // from the hub
  if (NavigationContext.QueryString.ContainsKey("keyString"))
  {
    var key = NavigationContext.QueryString["keyString"];

    // Find the media by your special key
    Song song = FindSongByKey(key);

    // Play the song normally

  }
}
```

By retrieving the key from the `NavigationContext`'s `QueryString` you can retrieve the `keyString` you added with the `MediaHistory-Item`. Once you have it, you can use the contents of the `keyString` to find the appropriate piece of media to play.

Working with the Camera

Although the `CameraCaptureTask` is an option for capturing an image using the phone's camera, it is not ideal when you need to have more control over the experience. For building your own camera applications, there is the `PhotoCamera` class that represents access to the photo hardware. When you need real-time access to the hardware, Windows Phone also supports Silverlight's raw hardware access APIs (including `Capture-Source`, `AudioSink`, and `VideoSink`).

Using the `PhotoCamera` Class

The `PhotoCamera` class represents the phone's camera and allows you to easily build photo-taking applications with it. To get started you will need an instance of the `PhotoCamera` class. You can use that class as the source of a `VideoBrush` to paint the "viewfinder" in your application, like so:

```
public partial class MainPage : PhoneApplicationPage
{
  PhotoCamera theCamera = null;

  protected override void OnNavigatedTo(NavigationEventArgs e)
  {
    theCamera = new PhotoCamera();

    // Set the Camera as the source for the VideoBrush
    previewBrush.SetSource(theCamera);

    ...
```

Using the `PhotoCamera` as the source of the `VideoBrush` causes a real-time image to be painted wherever you're using a `VideoBrush`. Once you have the camera created and showing up on the page, you need to wire up the camera's functionality in order to use it. The first thing to do in wiring the camera is to handle the `Initialized` event. This event is important to handle, as many of the camera settings (e.g., setting the focus or the flash) are not available until this event is fired. For example, to set the flash for red-eye reduction, handle the `Initialized` event and then set the flash like so:

```
// Some settings require camera to be initialized first
theCamera.Initialized += (s, a) =>
  {
    if (a.Succeeded)
    {
      // Taking Portraits
      if (theCamera.IsFlashModeSupported(FlashMode.RedEyeReduction))
      {
        theCamera.FlashMode = FlashMode.RedEyeReduction;
      }
    }
  };
```

The `Initialized` event argument includes a property to tell you whether it was successful (the `Succeeded` property). You should check this first to be certain that you can actually modify the `PhotoCamera`'s properties. In this case, we ask the camera if the red-eye reduction flash is supported on this phone, and if so, we can specify that is the flash mode we want.

In order to take a photo, you can use the `PhotoCamera`'s `Capture-Image` method. This tells the camera to take the photo and then fires events for both a full-size and a thumbnail version of the image. Typically, this is in response to some event such as a button click:

```
private void shutter_Click(object sender, RoutedEventArgs e)
{
  theCamera.CaptureImage();
}
```

Once the image has been captured, the `PhotoCamera` class can raise the `CaptureImageAvailable` and `CaptureThumbnailAvailable` events. These events include the actual stream that contains the image, so you can handle these events and manipulate the resultant photo. For example, you can take the full image and save it to the media library (as shown earlier in this chapter):

```
// Photo was captured
theCamera.CaptureImageAvailable += (s,a) =>
  {
     // Save picture to the device media library.
     MediaLibrary library = new MediaLibrary();
     library.SavePictureToCameraRoll("SomeFile.jpg", a.ImageStream);
  };
```

The `CaptureImageAvailable` event argument includes an `Image-Stream` property that contains the raw photo. You can manipulate this stream in any way you want, though saving it to the phone is more typical of what a user might expect.

Like you saw earlier, the `PhotoCamera` class offers several settings, including

- Flash mode
- Picture resolution

In addition, you can control the focus by using the `Focus` and `Focus-AtPoint` methods. These methods allow you to start the auto-focus functionality. You need to test to see if these capabilities are available:

```
private void focus_Click(object sender, RoutedEventArgs e)
{
   // AutoFocus the Camera
   if (theCamera.IsFocusSupported)
   {
      theCamera.Focus();
   }
}
```

Focusing at a specified point works the same way, but you need to specify a place to center on. The values should be between 0.0 and 1.1, so focus at whatever is in the center of the viewfinder:

```
private void focusAtPoint_Click(object sender, RoutedEventArgs e)
{
   // Focus in the center of the image
   if (theCamera.IsFocusAtPointSupported)
   {
      theCamera.FocusAtPoint(.5, .5);
   }
}
```

Before taking a photo, you will want to be sure the focusing is complete. To do this you can handle the `AutoFocusComplete` event:

```
bool _isFocusComplete = true;
...
theCamera.AutoFocusCompleted += (s, a) =>
  {
    _isFocusComplete = true;
  };
```

By keeping the state of whether your application is currently focusing, you can control when a picture is taken by flipping this flag during focus and shutter operations:

```
private void focus_Click(object sender, RoutedEventArgs e)
{
  // AutoFocus the Camera
  if (theCamera.IsFocusSupported)
  {
    _isFocusComplete = false;
    theCamera.Focus();
  }
}

private void shutter_Click(object sender, RoutedEventArgs e)
{
  if (_isFocusComplete) theCamera.CaptureImage();
}
```

You may want to use the hardware camera shutter to handle focus and shutter functions like the built-in camera application. The `CameraButtons` class gives you that capability. This class supports three events.

- **ShutterKeyPressed:** This occurs when the user fully presses the phone's dedicated camera button.
- **ShutterKeyHalfPressed:** This occurs when the user partially presses the phone's dedicated camera button. (It is usually used to start a focus operation.)
- **ShutterKeyReleased:** This occurs when the user releases the phone's dedicated camera button after fully pressing the button. (For example, it does not occur when the user half-presses the camera button.)

To allow the hardware button to control the camera app, you could handle these events as necessary:

```
// Wire up shutter button too (instead of UI control)
CameraButtons.ShutterKeyPressed += (s, a) => TakePicture();
CameraButtons.ShutterKeyHalfPressed += (s, a) => theCamera.Focus();
```

Finally, the `PhotoCamera` class gives you raw access to the preview buffer if you need to manipulate what the user sees in the virtual viewfinder. You would do this via the `GetPreviewBufferXXX` methods. You can get the buffer as 32-bit ARGB values, YCbCr, or just luminance data. You can see how this works here:

```
int[] buffer = new int[640 * 480];
theCamera.GetPreviewBufferArgb32(buffer);
// Manipulate the preview and show on the screen
```

The buffer you receive from these methods would have to be used to create a bitmap to show to the user. Often it is easier to just use the raw camera APIs (shown next) to accomplish this sort of real-time manipulation of the camera's feed.

■ Camera Capabilities

In order to use the `PhotoCamera` class (and other associated classes), you will need to ensure that your application includes the ID_CAP_ISV_CAMERA capability in the WMAppManifest.xml file.

Raw Hardware Access

The Windows Phone SDK includes a couple of key classes from Silverlight 4 that support low-level access to the camera and microphone hardware. Although it is much easier to use the `PhotoCamera` class, if you need to capture video and/or audio and manipulate it in real time these APIs are the best tool for the job.

The starting point to the raw camera API is the `CaptureSource` class. The `CaptureSource` class allows you to have access to video input on the phone. Like the `PhotoCamera` class, you can use the `CaptureSource` class as the source for a `VideoBrush`. This way, you can show the raw input from the camera in your application:

```
CaptureSource _src = new CaptureSource();

protected override void OnNavigatedTo(NavigationEventArgs e)
{
  // Show the preview
  previewBrush.SetSource(_src);
}
```

To enable the video, you need to enable the `CaptureSource` by using the `Start` method:

```
private void camButton_Click(object sender, RoutedEventArgs e)
{
  _src.Start();
}
```

■ Silverlight Developers

Unlike desktop Silverlight, you can start the `CaptureSource` without asking for permission with the `CaptureDeviceConfiguration` class.

At this point you only have the camera showing up in your own application. In order to deal with live input from the camera and microphone you have to create special classes called **sinks.** To retrieve video you need a `VideoSink` class; for audio you need an `AudioSink` class. These are abstract classes that you must derive from in order to retrieve real-time data from the hardware. They have a small number of abstract methods (that you have to override). A skeleton `VideoSink` derived class would look like this:

```
public class MyVideoSink : VideoSink
{
  protected override void OnSample(long
sampleTimeInHundredNanoseconds,
    long frameDurationInHundredNanoseconds,
    byte[] sampleData)
  {
    // Encode or Save the stream
  }

  protected override void OnCaptureStarted()
  {
```

```
    // Handle Startup of Capture
  }

  protected override void OnCaptureStopped()
  {
    // Cleanup After Capture
  }

  VideoFormat _format = null;

  protected override void OnFormatChange(VideoFormat videoFormat)
  {
    // Store the Video Format
    _format = videoFormat;
  }
}
```

Here are the abstract methods you must override.

- **OnCaptureStarted:** This is where you can do any initialization necessary to prepare for data capture.
- **OnCaptureStopped:** This is where you can clean up or save after a capture is complete.
- **OnFormatChange:** This is where you would store the current format to determine how to consume the capture samples.
- **OnSample:** This is where most of the real work is accomplished in the sink. This is called periodically with a set of samples from the hardware.

Once you have a sink, you can specify the `CaptureSource` of the sink before you start capturing:

```
protected override void OnNavigatedTo(NavigationEventArgs e)
{
  // Show the preview
  previewBrush.SetSource(_src);

  // Capture the video
  _sink.CaptureSource = _src;
}
```

When you specify the `CaptureSource` your sink will be notified as the `CaptureSource` is manipulated by the user. In this way, you can have more than one audio and video sink per `CaptureSource`. Explaining how to manipulate raw video/audio samples is outside the scope of this book.

Since the raw camera API originated from Silverlight, it enables you to specify the camera on the device you want. As of this writing, all phones only have one camera. But since this API is going to be supported for new phones as well, the `CaptureDeviceConfiguration` class is also supposed to allow you to enumerate the different audio and video capture devices. In this way, you can specify the capture device for the `Capture-Source` to use, like so:

```
ICollection<VideoCaptureDevice> devices =
  CaptureDeviceConfiguration.GetAvailableVideoCaptureDevices();

// Pick the first device
_src.VideoCaptureDevice = devices.First();
```

The API supports multiple devices so that, in the future, if there are multiple video devices (e.g., a front-facing camera), your code will be able to pick which device to use. By default, the `CaptureSource` uses the "primary device," which is usually the standard camera, so working with this API isn't necessary with the current crop of phones.

The Clipboard API

Windows Phone supports a common clipboard much like Windows. Users can add items to the clipboard while working with `TextBox` controls, and the copied text will remain in the clipboard for other applications to use. You can also access the clipboard programmatically with the `Clipboard` class. The `Clipboard` class has three simple static methods: `SetText`, `ContainsText`, and `GetText`. You can set text to the clipboard programmatically by using the `SetText` method, like so:

```
Clipboard.SetText("Hello World");
```

You can test to see if there is data in the clipboard by calling the `Con-tainsText` method as well:

```
if (Clipboard.ContainsText())
{
  // ...
}
```

The last method (`GetText`) exists on the `Clipboard` class but is not available to Windows Phone applications. Calling `GetText` will throw a security exception. This means your application cannot read the clipboard; it can only test to see if it has text and push text onto the clipboard. The reasoning behind this is to prevent accidental leaking of user data to unauthorized applications. For desktop Silverlight, usually user approval is required to access the clipboard, but currently the decision is to just disallow it instead of introducing another pop-up confirmation that needs to be explained to the user.

Live Tiles

In Chapter 6, Developing for the Phone, you learned that the WMManifest. xml file contains the images for the icon used both on the list applications for the phone and on applications pinned to the home screen. When a user pins your application to his home screen you can update the tile. This is called a Live Tile. Figure 7.13 shows the layers that make up a Live Tile.

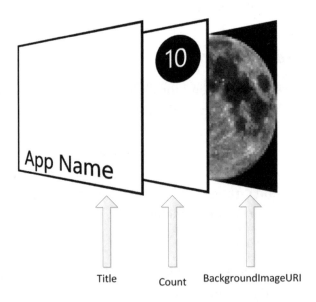

FIGURE 7.13 **Tile layers**

The layers of the tile correspond to the items in the WMManifest.xml file, as shown here:

```xml
<?xml version="1.0" encoding="utf-8"?>
<Deployment xmlns="..."
            AppPlatformVersion="7.1">
  <App ...>
    ...
    <Tokens>
      <PrimaryToken TokenID="FunWithTiles"
                    TaskName="_default">
        <TemplateType5>
          <BackgroundImageURI IsRelative="true"
                              IsResource="false">
          Background.png
          </BackgroundImageURI>
          <Count>0</Count>
          <Title>FunWithToast</Title>
        </TemplateType5>
      </PrimaryToken>
    </Tokens>
  </App>
</Deployment>
```

You probably have noticed that some of the built-in tiles, when pinned, change from time to time (like the email icons or the People hub). In the case of your application, the only things you can change are these three layers.

But you can do a lot more than specify these details for your Live Tiles. You can update the main tile as well as create new Live Tiles (that link to deeper parts of your application) and even create dual-sided Live Tiles. Let's look at each of these separately.

Main Live Tile

Your application has a single main Live Tile that is added to the home screen when a user manually pins your application. You can get at this Live Tile (whether it has been pinned or not) by accessing the `ShellTile.ActiveTiles` property. This property returns an enumerable list of the tiles in your application:

```csharp
IEnumerable<ShellTile> tiles = ShellTile.ActiveTiles;

// Get the default tile
var tile = tiles.First();
```

The `ShellTile` class represents a tile for the home screen. The `Active-Tiles` property is a collection of these `ShellTile` objects. The first one is always the default tile. This class allows you to update the default tile. To update the default tile, you can create a data structure called `Standard-TileData` that contains the data with which to update the tile:

```
IEnumerable<ShellTile> tiles = ShellTile.ActiveTiles;

// Get the default tile
var tile = tiles.First();

// The new tile information to use to update the tile
var tileData = new StandardTileData()
{
  Title = "My Application",
  Count = 54,
  BackgroundImage = new Uri("moon.png", UriKind.Relative),
};

// Update the Tile
tile.Update(tileData);
```

Calling the `Update` method will update the tile with the new information. If the default tile (in this case) has not been pinned yet, when the application is pinned it will include this new set of information instead of the information contained in the WMAppManifest.xml file.

Secondary Tiles

You can also create all new tiles that are pinned to the home screen for the user. The purpose of these secondary tiles is to allow you to have tiles that open deeper parts of your application. For example, imagine you have an application that shows the user RSS feeds. You may create a Live Tile that goes to a specific RSS feed instead of the default view that a new launching of the application would go to. This allows you to deep-link into your application. To do this, you use the same `StandardTileData` structure that was used to update a tile, but in this case you can create a new tile using the `ShellTile.Create` method:

```
var newTile = new StandardTileData()
{
  Title = "Favorite RSS Feed",
  BackgroundImage = new Uri("moon.png", UriKind.Relative)
};
```

```
var uri = "/RssView.xaml?feedname=favoriteFeedName";

ShellTile.Create(new Uri(uri, UriKind.Relative), newTile);
```

You should notice that the `ShellTile.Create` method requires a URI to be specified. This URI is used as the location from which to launch the navigation in your application. This is how deep linking works. You can see in this example the URI is going to a view we have called RssView. xaml and I am passing in a query string to help that view know how to show the data the Live Tile is specifying. When `ShellTile.Create` is called, it takes the user to the home screen and scrolls to the new tile (to allow the user to move it on the home screen if he desires). This behavior is not overridable. The purpose of going to the home screen is to prevent applications from hiding from the user the fact that they are adding new tiles. Since it deactivates the application, if the user did not want the Live Tile added he could just delete it and know that the application is nefarious and not launch it again.

You can find your own Live Tiles by interrogating the `NavigationUri` property of the `ShellTile` class (in the `ActiveTile` collection) like so:

```
var myTile = ShellTile.ActiveTiles
                      .Where(t => t.NavigationUri
                                   .OriginalString
                                   .Contains("RssView.xaml"))
                      .FirstOrDefault();

// If it was found
if (myTile != null)
{
  // ...
}
```

Once you find the appropriate tile, you can update it like you did the default tile:

```
// Set up new tile data
var tileUpdateData = new StandardTileData()
{
  Title = "Favorite RSS Feed",
  BackgroundImage = new Uri("moon.png", UriKind.Relative),
  Count = _stories.Count() // Some computed value
};
```

```
// Update the tile now
myTile.Update(tileUpdateData);
```

Typically you will want to change the tile with some information for the user to see that he should revisit your application (like updating the `Count` property).

If you need to change something else (like the `NavigationUri`), you need to delete the tile and re-create it. The `ShellTile` class allows you to delete a Live Tile (as long as it is not the default tile):

```
myTile.Delete();
```

On the whole, secondary tiles can be a very powerful feature, but the user will not necessarily tolerate a lot of tiles from a single application, so using this functionality judiciously is encouraged!

Dual-Sided Live Tiles

Live Tiles also can have a "back" side. The back of the tile is occasionally shown to the user, and is described using a background, title, and content. The background and title are the same as they are on the front of the tile. The content is essentially a short string. To specify the back of the tile you can use the `StandardTileData` structure, but you will set the additional properties for the back:

```
var newTile = new StandardTileData()
{
  Title = "Favorite RSS Feed",
  BackgroundImage = new Uri("moon.png", UriKind.Relative),
  BackBackgroundImage = new Uri("moonback.png", UriKind.Relative),
  BackContent = _stories.First().Title,
  BackTitle = "Feed Updated!"
};

var uri = "/RssView.xaml?feedname=favoriteFeedName";
ShellTile.Create(new Uri(uri, UriKind.Relative), newTile);
```

Remember that the back of the tile will only be shown occasionally (when it is shown depends on the operating system), so you want only additive information on the back. If you have new information for the user to encourage him to run the application, it should be on the front. Since the back data is in the `StandardTileData`, you can set it both when creating and when updating a tile.

> ## ▪▪ Looking Forward
>
> You can also update tiles using push notifications or with background agents. In Chapter 9, Multitasking, we will discuss how to use agents to update tiles. In Chapter 10, Services, you will learn how to use push notifications to update tiles.

Location APIs

Your application can determine where in the world the phone is at any moment. This is the same technology your GPS device uses to determine how to give you driving directions. The phones are required to have Assisted Global Positioning System (A-GPS) hardware. This is different from simple GPS location technology as the "Assisted" part is very important. Typical GPS devices rely on being able to locate three orbiting satellites in order to triangulate your location. Sometimes these systems were hurt by the lag in "syncing" with the three satellites, and often structures (such as buildings or bridges) made GPS unreliable. Assisted GPS improves this by using GPS to give you high-precision geolocation when possible, but can also fall back to use other ways of determining your location including cell-phone tower triangulation[2] and the WiFi Positioning System.[3]

Location Permission

In order to create a location-aware application, you must use the A-GPS on the phone. To access that device, you need to include the `System.Device` assembly from the Windows Phone SDK 7.1. This assembly will give you access to several classes to identify the current location of the phone. But before you can access the location information, you must find out if your application is allowed to access that information.

When your application is installed, the Marketplace specifically asks the user if he wants to allow your application to gather location information. Before your application can use location information, it must use the `GeoLocationWatcher` class (in the `System.Device.Location` namespace) to check whether the user has granted permission:

---

2. http://en.wikipedia.org/wiki/Mobile_phone_tracking
3. http://en.wikipedia.org/wiki/Wi-Fi_Positioning_System

```
// using System.Device.Location;
// Create the Watcher (and clean it up when done)
using (var watcher = new GeoCoordinateWatcher())
{
  if (watcher.Permission == GeoPositionPermission.Granted)
  {
    // Find the location
  }
}
```

The `Permission` property of the `GeoCoordinateWatcher` class will tell your application whether it is allowed to use location information. If permission is denied and you attempt to use location information, the class will throw an exception. In addition to checking for permission, you must allow the user to disable this capability in your application. This is typically accomplished in a "Settings" page or other mechanism. In Chapter 11, The Marketplace, we will discuss this requirement (and others) that your application must implement for certain phone features.

Accessing Location Information

The A-GPS on the phone allows you to retrieve location information without having to work with the sensor directly. You can use the `GeoCoordinate-Watcher` class to retrieve location information. When you create the `Geo-CoordinateWatcher` class you can optionally specify how accurate you need the geolocation information to be:

```
// Normal Accuracy
using (var watcher = new GeoCoordinateWatcher())
{
  // ...
}

// High Accuracy
using (var accurateWatcher =
  new GeoCoordinateWatcher(GeoPositionAccuracy.High))
{
  // ...
}
```

The level of accuracy impacts not only how accurate the information is, but also how fast you can get the information back. With the default level of accuracy, the `GeoCoordinateWatcher` class will return once it can get a location instead of waiting for high-accuracy information.

When developing your application, you need to also determine how often you need the geolocation information. Retrieving geolocation information can be a drain on the battery, so deciding whether you need a one-time location or whether you need to constantly monitor the change of location should be part of your application design. Let's look at ways to do both.

One-time Geolocation

To perform a one-time retrieval of the location of the phone, you can handle the GeoCoordinateWatcher class's StatusChanged event. This event will fire when the status of the location service changes. You can simply look for a status of "Ready" to retrieve the location directly from the GeoCoordinateWatcher class, like so:

```
public partial class MainPage : PhoneApplicationPage
{
  // Normal Accuracy
  GeoCoordinateWatcher _watcher = new GeoCoordinateWatcher();

  // Constructor
  public MainPage()
  {
    InitializeComponent();

    // For one-time query, just use StatusChanged
    _watcher.StatusChanged += _watcher_StatusChanged;

    // Start looking once the page has loaded
    Loaded += new RoutedEventHandler(MainPage_Loaded);

  }
```

Once the event has been registered, you would start the GeoCoordinateWatcher class in the Loaded event:

```
void MainPage_Loaded(object sender, RoutedEventArgs e)
{
  // Assuming the user gave permission, start the watcher
  if (_watcher.Permission == GeoPositionPermission.Granted)
  {
    _watcher.Start();
  }
}
```

You can query the `GeoCoordinateWatcher` class to see if the user gave permission to use the location information. If permission has been granted, you can start the watcher and wait for the `StatusChanged` event to fire:

```
void _watcher_StatusChanged(object sender,
                              GeoPositionStatusChangedEventArgs e)
{
  // If the status is ready, we can get the information
  if (e.Status == GeoPositionStatus.Ready)
  {
    // Show the Location in the UI
    DataContext = _watcher.Position.Location;

    // using the watcher can drain battery,
    // so turn it off when you're not
    // using it (and Dispose it too)
    _watcher.Stop();
    _watcher.Dispose();
  }
}
```

The `StatusChanged` event returns the status of the location information. This can be one of four statuses.

- **Ready:** Location information is available.
- **Initializing:** Location information should be available soon. The device is initializing.
- **NoData:** The device is working correctly but no data is available. This could be because none of the different location information is available (no GPS, no cell towers, and no Wi-Fi).
- **Disabled:** Location information was disabled for this application. This is usually if the user was asked and rejected the permission to let your application use location information.

For this one-time use, the `Ready` status is all you need. Once you get a `Ready` state, you can get the position directly from the `GeoCoordinate-Watcher` class. Since this is a one-time retrieval of information, you should turn off the watcher when you're done to prevent any battery drain from leaving the A-GPS device turned on.

The information retrieved from the `GeoCoordinateWatcher` class is a `GeoCoordinate` object. The `GeoCoordinate` class contains properties that give you the following information about the geolocation of the phone.

- **Latitude:** This is the current latitude of the phone (location on the surface of the Earth).

- **Longitude:** This is the current longitude of the phone (location on the surface of the Earth).

- **IsUnknown:** This says whether the coordinate is invalid or unknown.

- **Speed:** This is an approximate speed of the user in meters per second.

- **Course:** This is an approximate heading as related to true north (0–360 degrees).

- **Altitude:** This is the approximate altitude above sea level.

- **HorizontalAccuracy:** This is the distance of uncertainty in this location information (in meters).

- **VerticalAccuracy:** This is the distance of uncertainty in this location information (in meters).

Tracking Geolocation Changes

For applications that require continuous monitoring of the user's movements, you can monitor the geolocation changes as they occur. Like many of the devices on the phone, using the geolocation functionality for long periods may adversely impact battery life. To combat this you will want to be judicious in your use of this feature.

Continuously monitoring the geolocation of the phone is similar to a one-time retrieval of location information, but the method structuring the code is a little different. You should start by creating a `GeoCoordinate-Watcher` like before, but this time you will want to register for both the `StatusChanged` and the `PositionChanged` events:

```
public partial class MainPage : PhoneApplicationPage
{
  // Normal Accuracy
  GeoCoordinateWatcher _watcher = new GeoCoordinateWatcher();
```

```
// Constructor
public MainPage()
{
  InitializeComponent();

  // To monitor we should register for both events
  _watcher.StatusChanged += _watcher_StatusChanged;
  _watcher.PositionChanged += _watcher_PositionChanged;

}
```

Like before, you'll want to check to see if you have permission to use the location services before you start the watcher:

```
private void startButton_Click(object sender, RoutedEventArgs e)
{
  // Assuming the user gave permission, start the watcher
  if (_watcher.Permission == GeoPositionPermission.Granted)
  {
    _watcher.Start();
  }
}
```

It is the event handlers that are different in this case. First, in the StatusChanged event you will want to tell the user what is happening based on the statuses that are returned, like so:

```
void _watcher_StatusChanged(object sender,
                            GeoPositionStatusChangedEventArgs e)
{
  switch (e.Status)
  {
    // Tell the user the location information is coming
    case GeoPositionStatus.Initializing:
      statusMessage.Text = "Location Initializing";
      break;

    // Alert the user that data isn't available currently
    case GeoPositionStatus.NoData:
      statusMessage.Text = "No location available...try again later";
      break;

    // If status is Ready, then PositionChanged fired correctly
    case GeoPositionStatus.Ready:
      statusMessage.Text = "Receiving Location Information";
      break;
```

```
    // If Disabled, tell the user
    case GeoPositionStatus.Disabled:
      statusMessage.Text = "Location Information is disabled";
      break;
  }
}
```

Because these statuses can change while your application is running, you should continue to monitor the `GeoPositionStatus` values and keep the user apprised of the validity of this information. Note that, unlike the one-time retrieval, we do not get the status information from the watcher when the `Ready` status fires. This is because the other event (`PositionChanged`) will be fired as the location changes:

```
void _watcher_PositionChanged(object sender,
                      GeoPositionChangedEventArgs<GeoCoordinate> e)
{
  // Use location information
  DataContext = e.Position.Location;
}
```

The `Ready` status will only fire once, even though the location is changing, so the `PositionChanged` event is where you will actually use the location information as it changes. You should support stopping the location information as well, which you can do with the `Stop` method of the `GeoCoordinateWatcher` class:

```
private void stopButton_Click(object sender, RoutedEventArgs e)
{
  // Stop the watcher
  _watcher.Stop();
}
```

■ Turning Coordinates into Addresses

Now that you have basic coordinates of where the phone is on the surface of the Earth, you will want to turn that into something the user can actually use. This requires use of certain services that we will cover in Chapter 10 , Services.

Emulating Location Information

When using the emulator, you can emulate location (e.g., A-GPS) information for your application for testing. To accomplish this, you need to open the Additional Tools sidebar of the emulator (like you did when you looked at the accelerometer emulation earlier in this chapter), as shown in Figure 7.14.

The Additional Tools dialog includes two tabs. For location information, you'll want to click on the Location tab, as shown in Figure 7.15.

The Location section of the Additional Tools dialog includes several pieces of key functionality. You can see the basic layout of the location user interface in Figure 7.16.

The main part of the window (labeled #1) is the map. You can scroll around and double-click/scroll-wheel to zoom on the map. In the upper-left-hand corner of the window is a search box (#2) to allow you to find a location on the map. Typing in a name, zip code, or other description will focus it on the map. Next to the search box are zoom tools to help you find the right level of detail. Next, the Live button (#3) is a toggle to enable live mode. In this mode, the emulator will use the center of the map as the current location for the phone.

FIGURE 7.14 Opening the emulator's Additional Tools sidebar

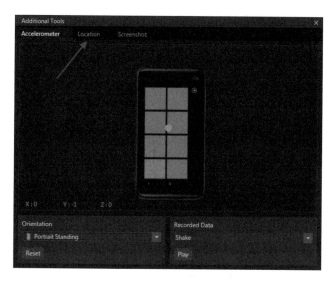

FIGURE 7.15 Selecting the Location tab

FIGURE 7.16 Location tab of the Additional Tools dialog

Finally, you can also specify map pins if the pin button (#4) is selected. This way, you can create routes (the pins are sequentially numbered as you click them) so that you can see what happens as you are simulating a set of locations. With the pin button pressed, every time you click on the map it will create a waypoint, as shown in Figure 7.17.

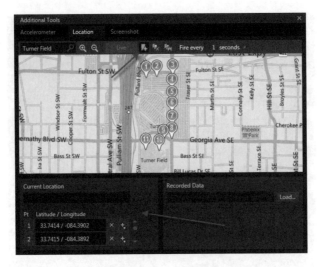

Figure 7.17 Using pins to create waypoints

If you unselect Live, you'll be able to press the play key in the upper-right-hand corner to actually play through the set of waypoints. You should notice that the lower-left-hand corner of the dialog shows you all the waypoints and allows you to reorder and delete individual waypoints. You can also clear all the waypoints by clicking the "delete all points" button to the right of the pin button.

Finally, you can take your waypoints and timings and save them for replay. To do this, click on the save button shown in Figure 7.18.

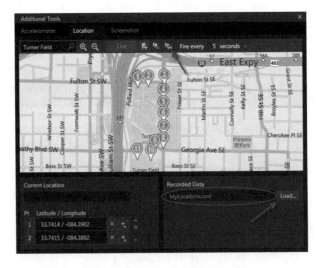

Figure 7.18 Saving recorded data

You can reload saved data by using the Load button in the lower-right-hand part of the dialog. This way you can create complex location information and use it to test your application. By using these features you can simulate location information in the emulator and enable testing scenarios that would be impossible in real life.

Where Are We?

This is where phone development is really different from almost every other type of application design: integration with the phone. In this chapter you saw how to access information from the device, store information on the device, and interact with the core phone functions such as the camera and the phone functions themselves. Making your application a real phone application will require use of these types of functionality; otherwise, it is just an app on a handheld device.

8

Databases and Storage

A s you're writing your new, world-changing application you will sometimes need to store information directly on the phone. This information may be simple files, or caches to serialized versions of data, or even a full-fledged database. Since the storage is on the phone, it is important to understand all your options so that you can make the right decision about where to store information as well as take the fewest system resources necessary to do the job. In this chapter you will learn how to make those decisions as well as implement them.

Storing Data

Apps need data. Data needs to be persistent. That's the big picture. You need to be able to store data on a phone for use by your application. That data needs to remain on the phone between invocations of your application as well as through reboots. While Windows Phone supports this in several ways, you will want to understand the different facilities to best decide how to accomplish this.

The two main ways to store data are through isolated storage and a local database. **Isolated storage** allows you to store data as files in a virtualized file system that only your application has access to. In addition to storing files, you can also store information in a **local database** that provides you with a way to store data and query the data for retrieval. From those broad strokes it might seem that storing data in a database is the way to go. This is often the most common approach for developers transitioning to the phone. The problem is that you need to determine if you need a database.

Using the phone's database requires more resources than simply storing files. It may seem easier, but that simplicity comes at a cost. The reality is that many applications do not need the query and update facilities of a full-blown database system. Often what they really need is to simply store data for the next time the application is started.

Isolated Storage

As stated earlier, one of the options is to use the file system on the phone. Each application is allocated a separate private part of the internal file system for its use. This means any files you store in this area (called isolated storage) is for your application's use only. It is not accessible by any other application (though the operating system has access to these files). When your application is installed, this area of the file system is allocated. Likewise, when your application is uninstalled, any files that are in isolated storage are lost.

Accessing isolated storage on the phone is similar to accessing it in other versions of Silverlight (or even standard .NET). You use the isolated storage classes to open files or directories and use the `System.IO` stack to read, edit, and write files. Let's get started and see how to use the isolated storage classes.

■ Isolated Storage Quota

While in the desktop version of Silverlight isolated storage has a specified quota to limit how much data can be stored on the device, the phone is different. The quota is not factored in on the phone; only the device's free space matters. The APIs include properties and methods for the quota on the phone, but you should never use these as the quota will always be larger than the available space on the device.

To work with isolated storage you have to first retrieve an object that will represent the gateway to the phone's storage. You can do this with the `IsolatedStorageFile` object (in the `System.IO.IsolatedStorage` namespace). This name is a little misleading, because it's not a file but an object that represents the "store" for your application. You retrieve the object by calling the `IsolatedStorageFile`'s `GetUserStoreFor-Application` static method:

```
using (IsolatedStorageFile store =
        IsolatedStorageFile.GetUserStoreForApplication())
{
  // Use the store
}
```

The `IsolatedStorageFile` class supports `IDisposable`, so you should wrap it in a `using` statement (as shown in the example) to ensure that it releases all the resources once it is done.

> **■ Performance Tip**
>
> The `IsolatedStorageFile` object should be disposed when you are done accessing it. But since accessing the object is expensive, you may want to cache this object for multiple uses. Wrapping the object with a `using` block is useful unless you are going to be saving multiple files. In that case, you should reuse the `IsolatedStorageFile` object.

The `IsolatedStorageFile` class supports methods for creating and opening files and directories. You can think of the store as the starting point in a directory structure in which to store your data. For example, you can create a new file by using the `CreateFile` method, like so:

```
using (IsolatedStorageFile store =
        IsolatedStorageFile.GetUserStoreForApplication())
{
  using (IsolatedStorageFileStream file =
          store.CreateFile("settings.txt"))
  {
    StreamWriter writer = new StreamWriter(file);
    writer.WriteLine("FavoriteColor=Blue");
    writer.Close();
  }
}
```

The `IsolatedStorageFileStream` that is returned from the `CreateFile` method can be treated as a simple `Stream`. So writing to the `Stream` using standard .NET techniques works just fine (as shown in the example with a `StreamWriter`). You can do the same thing to read a file with the `OpenFile` method:

```
using (IsolatedStorageFile store =
        IsolatedStorageFile.GetUserStoreForApplication())
{
  using (IsolatedStorageFileStream file =
          store.OpenFile("settings.txt", FileMode.Open))
  {
    StreamReader reader = new StreamReader(file);
    var line = reader.ReadToEnd();
    reader.Close();
  }
}
```

■ Isolated Storage and the Emulator

When you are working with isolated storage in the emulator, the data in isolated storage will persist over multiple invocations of the application but will be cleared under two circumstances: restarting the emulator and performing a "clean" build of your Visual Studio project. A clean build from Visual Studio will cause the application in the emulator to completely clear out isolated storage. This clean build happens during a "rebuild" as well, but not during standard builds or simple running of your application.

Serialization

In general, creating your own files and directories will help you store the information you need, but inventing your file formats is usually unnecessary. There are three common methods for storing information in a structured way using serialization: XML serialization, JSON serialization, and isolated storage settings.

XML Serialization

XML is a common approach to formatting data on the phone. .NET provides built-in support for XML serialization using the `XmlSerializer` class. This class allows you to take a graph of managed objects and create an XML version. Additionally (and crucially), you can read that XML back into memory to re-create the object graph.

To get started writing an object graph as XML, you will need to know about the class you want to serialize. For the examples here, assume you have something like a class that holds user preferences like so:

```
public class UserPreferences
{
  public DateTime LastAccessed { get; set; }
  public string FirstName { get; set; }
  public string LastName { get; set; }
  public bool UsePushNotifications { get; set; }
}
```

Standard XML serialization requires certain constructors (empty constructors) to work. If you don't create a constructor explicitly, the compiler will create an empty constructor implicitly. Because XML serialization requires an empty constructor, you can rely on the one the compiler created (as in the example above), but if you add an explicit constructor you need to make sure you also include a constructor without any parameters (i.e., an empty constructor). Now that we have something to serialize, let's go ahead and store the data.

First you will need to add the `System.Xml.Serialization` assembly to your project. Then you can use the `XmlSerialization` class (located in the `System.Xml.Serialization` namespace) by creating an instance of it (passing in the type that you want to serialize):

```
using (var store = IsolatedStorageFile.GetUserStoreForApplication())
using (var file = store.CreateFile("myfile.xml"))
{
  XmlSerializer ser = new XmlSerializer(typeof(UserPreferences));
}
```

This creates a serializer for this type. The instance of the class can serialize or deserialize your type (and associated types) to XML. To serialize into XML, simply call the `Serialize` method:

```
var someInstance = new UserPreferences();

using (var store = IsolatedStorageFile.GetUserStoreForApplication())
using (var file = store.CreateFile("myfile.xml"))
{
  XmlSerializer ser = new XmlSerializer(typeof(UserPreferences));
```

```
    // Serializes the instance into the newly created file
    ser.Serialize(file, someInstance);
}
```

This stores the instance of the `UserPreferences` class into the file that was created in isolated storage. To read that file back, you can reverse the process by using the `Deserialize` method like so:

```
UserPreferences someInstance = null;

using (var store = IsolatedStorageFile.GetUserStoreForApplication())
using (var file = store.OpenFile("myfile.xml", FileMode.Open))
{
  XmlSerializer ser = new XmlSerializer(typeof(UserPreferences));

  // Re-creates an instance of the UserPreferences
  // from the Serialized data
  someInstance = (UserPreferences)ser.Deserialize(file);
}
```

By using the `Deserialize` method, you can re-create the saved data and instantiate a new instance of the data with the serialized data. The serialization class also supports a `CanDeserialize` method, so you can test to see whether the stream contains serializable data that is compatible with the type:

```
UserPreferences someInstance = null;

using (var store = IsolatedStorageFile.GetUserStoreForApplication())
using (var file = store.OpenFile("myfile.xml", FileMode.Open))
{
  XmlSerializer ser = new XmlSerializer(typeof(UserPreferences));

  // Use an XmlReader to allow us to test for serializability
  var reader = XmlReader.Create(file);

  // Test to see if the reader contains serializable data
  if (ser.CanDeserialize(reader))
  {
    // Re-creates an instance of the UserPreferences
    // from the Serialized data
    someInstance = (UserPreferences)ser.Deserialize(reader);
  }
}
```

To test whether the data contains serializable data, you need to create an `XmlReader` object that contains the file contents (as a stream) using the

`XmlReader.Create` method. Once you have an `XmlReader`, you can use the serializer's `CanDeserialize` method. If you are going to wrap the file in an `XmlReader`, you should also send the `reader` object into the `Deserialize` method (as this is more efficient than the `XmlSerializer` creating a new reader internally).

> ### ▪ XML Serialization Choices
>
> Although you can see how to serialize your objects using the built-in XML serialization classes, you can also use third-party classes or even the `DataContractXmlSerialization` class from WCF to perform the serialization. The concepts are the same.

JSON Serialization

Although XML is the first choice of a lot of developers, it probably should not be today. The proliferation of JavaScript Object Notation (JSON) means that using the same format for local serialization and any Web- or REST-based interaction is commonplace. In addition, JSON tends to be smaller when serialized than XML. While size alone isn't a reason to choose JSON, smaller generally means faster, and that is a good reason to pick it.

In place of the `XmlSerializer` class, you can use the `DataContract-JsonSerializer` class. This class is part of the WCF (or Service Model) classes that are part of the Windows Phone SDK. This class lives in the `System.Runtime.Serialization` namespace but is implemented in the `System.ServiceModel.Web` assembly. This assembly is not included by default, so you need to add it manually.

To use the `DataContractJsonSerializer` class, you can create it by specifying the data type to store, like so:

```
UserPreferences someInstance = new UserPreferences();

using (var store = IsolatedStorageFile.GetUserStoreForApplication())
using (var file = store.CreateFile("myfile.json"))
{
  // Create the Serializer for our preferences class
  var serializer = new
    DataContractJsonSerializer(typeof(UserPreferences));
```

```
// Save the object as JSON
serializer.WriteObject(file, someInstance);

}
```

You can see here that the `DataContractJsonSerializer` class's `WriteObject` method is used to serialize the instance of the `User-Preferences` class as JSON. For deserializing the data back into an instance of the `UserPreferences` class, you would use the `ReadObject` method instead:

```
UserPreferences someInstance = null;

using (var store = IsolatedStorageFile.GetUserStoreForApplication())
using (var file = store.OpenFile("myfile.json", FileMode.Open))
{
  // Create the Serializer for our preferences class
  var serializer = new
    DataContractJsonSerializer(typeof(UserPreferences));

  // Load the object from JSON
  someInstance = (UserPreferences)serializer.ReadObject(file);
}
```

In this case, the serialization code simply used the `DataContact-JsonSerializer` class's `ReadObject` method to load a new instance of the `options` class by reading the JSON file that was earlier saved to storage.

Isolated Storage Settings

Sometimes it is just easier to let the system handle serialization for you. That's the concept behind isolated storage settings. The `IsolatedStorage-Settings` class represents access to a simple dictionary of things to store for the phone. This type of storage is useful for small pieces of information (e.g., application settings) that you need to store without the work of building a complete serialization scheme.

To get started you'll need to get the `settings` object for your application through a static accessor on `IsolatedStorageStatic` (in the `System.IO.IsolatedStorage` namespace) called `Application-Settings`, like so:

```
IsolatedStorageSettings settings =
  IsolatedStorageSettings.ApplicationSettings;
```

This class exposes a dictionary of properties that are automatically serialized to isolated storage for you. For example, if you want to store a setting when closing the application and read it back when launching the application you can do this with the `IsolatedStorageSettings` class like so:

```
Color _favoriteColor = Colors.Blue;
const string COLORKEY = "FavoriteColor";

void Application_Launching(object sender, LaunchingEventArgs e)
{
  if (IsolatedStorageSettings.ApplicationSettings.Contains(COLORKEY))
  {
    _favoriteColor =
      (Color)IsolatedStorageSettings.ApplicationSettings[COLORKEY];
  }
}

void Application_Closing(object sender, ClosingEventArgs e)
{
  IsolatedStorageSettings.ApplicationSettings[COLORKEY] =
    _favoriteColor;
}
```

By using the `IsolatedStorageSettings` class's `Application-Settings` property, you can set and get simple values. You can see that during the launch of the application, the code first checks to see if the key is in the `ApplicationSettings` (to be sure it's been saved at least once) and if so, it loads the value from isolated storage.

The `IsolatedStorageSettings` class does work well with simple types, but as long as the data you want to store is compatible with XML serialization (like our earlier example), the settings file will support saving it as well:

```
UserPreferences _preferences = new UserPreferences();
const string PREFKEY = "USERPREFS";

void Application_Launching(object sender, LaunchingEventArgs e)
{
  if (IsolatedStorageSettings.ApplicationSettings.Contains(PREFKEY))
  {
    _preferences = (UserPreferences)
      IsolatedStorageSettings.ApplicationSettings[PREFKEY];
  }
}
```

```
void Application_Closing(object sender, ClosingEventArgs e)
{
  IsolatedStorageSettings.ApplicationSettings[PREFKEY] =
    _preferences;
}
```

Local Databases

When you are building an application that needs data that must be queried and support smart updating, a local database is the best way to accomplish that. Windows Phone supports databases that exist directly on the phone. When building a Windows Phone application, you won't have access to the database directly; instead you can use a variant of LINQ to SQL married to a code-first approach to build a database to accomplish your database access. Let's walk through the meat of the functionality.

Getting Started

To get started you will need a database file. Under the covers the database is SQL Server Compact Edition, so you could just create an .sdf file for your project, but usually you will start by telling the database APIs to create the database for you.

> **■ Looking Forward**
>
> You can include local database files (SQL Server Compact Edition or .sdf files) with your application. We will cover this later in this chapter.

The first step is to have a class (or several) that represents the data you want to store. You can start with a simple class:

```
public class Game
{
  public string Name { get; set; }
  public DateTime? ReleaseDate { get; set; }
  public double? Price { get; set; }
}
```

This class holds some piece of data we want to be able to store in a database. Before we can store it in the database, we have to add attributes to tell LINQ to SQL that this describes a table:

```
[Table]
public class Game
{
    [Column]
    public string Name { get; set; }

    [Column]
    public DateTime? ReleaseDate { get; set; }

    [Column]
    public double? Price { get; set; }
}
```

By using these attributes, we are creating a class that represents the storage for a table in the database. Some of the column information is inferred (such as nullability in the `ReleaseDate` column). With this definition, we can read from the database, but before we can add or change data we need to define a primary key:

```
[Table]
public class Game
{
    [Column(IsPrimaryKey = true, IsDbGenerated = true)]
    public int Id { get; set; }

    [Column]
    public string Name { get; set; }

    [Column]
    public DateTime? ReleaseDate { get; set; }

    [Column]
    public double? Price { get; set; }
}
```

As you can see, the `Column` attribute has several properties that can be set to specify information about each column. In this case, the `Column` attribute specifies that the `Id` column is the primary key and that the key should be generated by the database. To support change tracking and writing to the database, you must have a primary key.

Like any other database engine, SQL Server Compact Edition allows you to improve query performance by adding your own indexes. You can do this by adding the `Index` attribute to the table classes:

```
[Table]
[Index(Name = "NameIndex", Columns = "Name", IsUnique = true)]
public class Game
{
  // ...
}
```

The `Index` attribute allows you to specify a name, a string containing the column names to be indexed, and optionally whether the index should also be a unique constraint. This attribute is used when you create or update the database. You can also specify an index on multiple columns by separating the column names with a comma:

```
[Table]
[Index(Name = "NameIndex", Columns = "Name", IsUnique = true)]
[Index(Columns = "ReleaseDate,IsPublished")]
public class Game : INotifyPropertyChanging, INotifyPropertyChanged
{
  // ...
}
```

At this point you have defined a simple table with two indexes and can move on to creating a data context class. This class will be your entry point to the database itself. It is a class that derives from the `DataContext` class, as shown here:

```
public class AppContext : DataContext
{
}
```

This class is responsible for exposing access to the database as well as handling change management. You will expose your "tables" as a public field:

```
public class AppContext : DataContext
{
  public Table<Game> Games;
}
```

The generic class wraps your table class to represent a queryable set of those objects. This way, your context class will not only give you access to the objects stored in the database, but also track them for you. The base class (DataContext) is where most of the magic happens. Since the DataContext class does not have an empty constructor, you'll also need to implement a constructor:

```
public class AppContext : DataContext
{
  public AppContext()
    : base("DataSource=isostore:/myapp.sdf;")
  {
  }

  public Table<Game> Games;
}
```

The typical call to the base class's constructor requires that you send it a connection string. For the phone, all this connection string requires is a description of where the database exists or where it will be created. You specify this by specifying a URI to where the database file belongs. The URI is a path to the file from either isolated storage or the application folder. To specify a file to exist (or be created) in isolated storage you use the isostore moniker like so:

isostore:/myapp.sdf

For a database that ships with your application (and will be delivered in the .xap file), you can use the appdata moniker as well. If you want to access data that resides in the application folder, you will only be able to read the database, not write to it. Later in this chapter you will see how to copy the database to the isostore folder if you need to write to an app-delivered database. To specify the location of the database in the application folder, you can use the appdata moniker just like the isostore moniker:

appdata:/myapp.sdf

The end of the URI should be a path and a file name to the actual file. The underlying database is SQL Server Compact Edition (i.e., SQL CE), so the file is an .sdf file. If you want your database file to be within a sub-folder, you can specify it in the URI with the folder name like so:

isostore:/**data/myapp.sdf**

Once you have created your data context class, you can create the database by calling the `CreateDatabase` method (as well as checking to see if it exists by calling `DatabaseExists`):

```
// Create the Context
var ctx = new AppContext();

// Create the Database if it doesn't exist
if (!ctx.DatabaseExists())
{
   ctx.CreateDatabase();
}
```

The context's table members allow you to perform CRUD[1] on the underlying data. For example, to create a new Game object in the database, you would just create an instance and add it to the Games member:

```
// Create a new game object
var game = new Game()
{
   Name = "Gears of War",
   Price = 39.99,
};

// Queue it as a change
ctx.Games.InsertOnSubmit(game);

// Submit all changes (inserts, updates and deletes)
ctx.SubmitChanges();
```

The new Game object can be passed to the Games member of the context through the InsertOnSubmit method to tell the context to save this the next time changes are submitted to the database. The SubmitChanges method will take any changes that have occurred since the creation of the context object (or since the last call to SubmitChanges) and batch them to the underlying database. Note that the new instance of Game didn't set the Id property. This is unnecessary as the Id property is marked not only as the primary key (which is required to support writing to the database) but also as database-generated. This means that when SubmitChanges is called, it will let the database generate the Id and update your object's ID to the database-generated one.

---

1. Create, Read, Update, and Delete

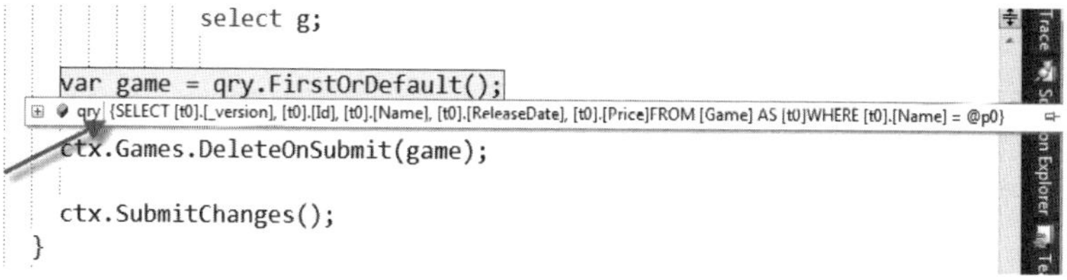

FIGURE 8.1 The SQL query

Querying the `Games` stored in the database takes the form of LINQ queries. So if you have created some data in the database, you can query it like so:

```
var qry = from g in ctx.Games
          where g.Price >= 49.99
          order by g.Name
          select g;

var results = qry.ToList();
```

This query will return a set of `Game` objects with the data directly from the database. This LINQ query is translated into a parameterized SQL query and executed against the local database for you when this code calls the `ToList` method. In fact, you can see the translated query in Visual Studio while debugging, as shown in Figure 8.1.

What may not be obvious is that if you change these objects, the context class tracks those changes for you. So if you change some data, calling the context's `SubmitChanges` method updates the database as well:

```
var qry = from g in ctx.Games
          where g.Name == "Gears of War"
          select g;

var game = qry.First();

game.Price = 34.99;

// Saves any changes to the game
ctx.SubmitChanges();
```

In addition, you can delete individual items using the table members on the context class by calling `DeleteOnSubmit` like so:

```
var qry = from g in ctx.Games
          where g.Name == "Gears of War"
          select g;

var game = qry.FirstOrDefault();

ctx.Games.DeleteOnSubmit(game);

// Saves any chances to the game
ctx.SubmitChanges();
```

You do need to retrieve the entities in order to delete them (unlike the full version of LINQ to SQL where you could execute arbitrary SQL). You can submit a deletion by calling `DeleteAllOnSubmit` and supplying a query:

```
var qry = from g in ctx.Games
          where g.Price > 100
          select g;

ctx.Games.DeleteAllOnSubmit(qry);

ctx.SubmitChanges();
```

The query in this example defines the items to be deleted in the database. It does not retrieve them in this place, but uses the query to define what items are to be deleted. Once all the items are marked for deletion, the call to `SubmitChanges` causes the deletion to happen (as well as any other changes that were detected).

By creating your table classes and a context class, you can access the database and perform all the queries and changes to the database that are necessary. Next let's look at additional database features you will probably want to consider as part of your phone application.

Optimizing the Context Class

Although the context class will track your objects, you can help the context class by ensuring that your table classes support the `INotifyProperty-Changing` and `INotifyPropertyChanged` interfaces. Implementing these interfaces has the additional benefit of assisting with data binding in Silverlight. Therefore, it is recommended that all your table classes support this interface, like so:

```
[Table]
public class Game : INotifyPropertyChanging, INotifyPropertyChanged
{
  // …

  public event PropertyChangingEventHandler PropertyChanging;

  public event PropertyChangedEventHandler PropertyChanged;

  void RaisePropertyChanged(string propName)
  {
    if (PropertyChanged != null)
    {
      PropertyChanged(this, new PropertyChangedEventArgs(propName));
    }
  }

  void RaisePropertyChanging(string propName)
  {
    if (PropertyChanging != null)
    {
      PropertyChanging(this,
                    new PropertyChangingEventArgs(propName));
    }
  }
}
```

Implementing both interfaces will add the `PropertyChanging` and `PropertyChanged` events to your class. As seen here, creating a simple helper method to raise these events is a common practice. Now that the interfaces are implemented, you have to use them. This involves calling the helper methods in each property setter. Our original `Game` class used automatic properties to expose the columns, but since we need to call the helper method we need standard properties:

```
[Table]
public class Game : INotifyPropertyChanged
{
  int _id;

  [Column(IsPrimaryKey = true, IsDbGenerated = true)]
  public int Id
  {
    get { return _id; }
    set
    {
      RaisePropertyChanging("Id");
```

```csharp
      _id = value;
      RaisePropertyChanged("Id");
    }
  }

  string _name;

  [Column]
  public string Name
  {
    get { return _name; }
    set
    {
      RaisePropertyChanging("Name");
      _name = value;
      RaisePropertyChanged("Name");
    }
  }

  DateTime? _releaseDate;

  [Column]
  public DateTime? ReleaseDate
  {
    get { return _releaseDate; }
    set
    {
      RaisePropertyChanging("ReleaseDate");
      _releaseDate = value;
      RaisePropertyChanged("ReleaseDate");
    }
  }

  double? _price;

  [Column]
  public double? Price
  {
    get { return _price; }
    set
    {
      RaisePropertyChanging("Price");
      _price = value;
      RaisePropertyChanged("Price");
    }
  }

  // ...
}
```

You should notice that each property now has a backing field member (e.g., _id for the Id property) and calls the RaisePropertyChanging and RaisePropertyChanged methods with the name of the property when the setter is called. By using these interfaces, the memory footprint of the context object is much smaller as it uses these interfaces to monitor changes.

In addition to these interfaces, you can improve the size of your update and delete queries by including a version member of your class:

```
[Table]
public class Game : INotifyPropertyChanging, INotifyPropertyChanged
{
  // ...

  [Column(IsVersion = true)]
  private Binary _version;
}
```

The version column (IsVersion = true) is optional but will improve the performance of change tracking when using database data. The version must be of type Binary from the System.Data.Linq namespace. It can be a private field (so it's not visible to users) but does need to be marked as IsVersion = true for LINQ to SQL to consider it the version column.

Performance Recommendation

It is recommended that your table classes support a primary key column and a version column, and that they implement the INotifyProperty-Changing and INotifyPropertyChanged interfaces in order to be as efficient as possible in your database access code.

Finally, if your database is only performing queries, you can tell the context class that you do not want to monitor any change management. You would accomplish this by setting the context class's ObjectTrack-ingEnabled property to false like so:

```
using (var ctx = new AppContext())
{
  ctx.ObjectTrackingEnabled = false;
```

```
var qry = from g in ctx.Games
          where g.Price < 19.99
          orderby g.ReleaseDate descending
          select g;

var results = qry.ToList();
}
```

By disabling change management, the context object will be much lighter weight. Also, because the context is not necessary for tracking the change, you can create it locally and dispose of it when the query is complete. Normally you would keep the context around for the lifetime of the page or application so that it can monitor and batch those changes back to the database, but since you are only reading from the database, the lifetime can be shortened if needed.

Associations

The data types for each property in the table classes you have seen have been simple types. The types have been simple because they need to be stored in the local database. In order to be stored in the local database they need to be convertible to database types (e.g., strings are stored as NVAR-CHARs). Even though you're going to be dealing with classes, you will still have to remember that it is a relational database underneath the covers. So when you need more structure, you will need associated tables (or associations).

For example, let's assume we have a second table class that holds information about the publisher of a game:

```
[Table]
public class Publisher :
  INotifyPropertyChanging, INotifyPropertyChanged
{
  int _id;

  [Column(IsPrimaryKey = true, IsDbGenerated = true)]
  public int Id
  {
    get { return _id; }
    set
    {
      RaisePropertyChanging("Id");
      _id = value;
```

```
      RaisePropertyChanged("Id");
    }
  }

  string _name;

  [Column]
  public string Name
  {
    get { return _name; }
    set
    {
      RaisePropertyChanging("Name");
      _name = value;
      RaisePropertyChanged("Name");
    }
  }

  string _website;

  [Column]
  public string Website
  {
    get { return _website; }
    set
    {
      RaisePropertyChanging("Website");
      _website = value;
      RaisePropertyChanged("Website");
    }
  }

  [Column(IsVersion = true)]
  private Binary _version;

  public event PropertyChangingEventHandler PropertyChanging;

  public event PropertyChangedEventHandler PropertyChanged;

  void RaisePropertyChanged(string propName)
  {
    if (PropertyChanged != null)
    {
      PropertyChanged(this, new PropertyChangedEventArgs(propName));
    }
  }

  void RaisePropertyChanging(string propName)
  {
```

```
    if (PropertyChanging != null)
    {
      PropertyChanging(this,
        new PropertyChangingEventArgs(propName));
    }
  }
}
```

This new class is implemented just like the Game class (since we want it to allow change management). To be able to save it in the database, we need to expose it on our context class as well as on a public field:

```
public class AppContext : DataContext
{
  public AppContext()
    : base("DataSource=isostore:/myapp.sdf;")
  {
  }

  public Table<Game> Games;

  public Table<Publisher> Publishers;
}
```

At this point you could create, edit, query, and delete both the Game and Publisher objects. But what you really want is to be able to relate the two objects to each other. That's where associations come in.

To add an association we need to start by having a column on the Game class that represents the publisher's primary key:

```
[Table]
public class Game : INotifyPropertyChanging, INotifyPropertyChanged
{
  // ...

  [Column]
  internal int _publisherId;

}
```

This new column is used to hold the ID of the related publisher for this particular game. The data is not public (it is internal in this case) because users of this class won't set this value explicitly. Instead, you will create a nonpublic member that will store an object called an EntityRef. The EntityRef class is a generic class that wraps a related entity:

```
[Table]
public class Game : INotifyPropertyChanging, INotifyPropertyChanged
{
  // ...

  [Column]
  internal int _publisherId;

  private EntityRef<Publisher> _publisher;
}
```

The `EntityRef` class is important here as it will also support lazy loading of the related entity so that large object graphs aren't loaded accidentally. But the real magic of linking the column and the `EntityRef` happens in the `public` property for the related entity:

```
[Table]
public class Game : INotifyPropertyChanging, INotifyPropertyChanged
{
  // ...

  [Column]
  internal int _publisherId;

  private EntityRef<Publisher> _publisher;

  [Association(IsForeignKey = true,
    Storage = "_publisher",
    ThisKey = "_publisherId",
    OtherKey = "Id")]
  public Publisher Publisher
  {
    get { return _publisher.Entity; }
    set
    {
      // Handle Change Management
      RaisePropertyChanging("Publisher");

      // Set the entity of the EntityRef
      _publisher.Entity = value;

      if (value != null)
      {
        // Set the foreign key too
        _publisherId = value.Id;
      }
```

```
        // Handle Change Management
        RaisePropertyChanged("Publisher");
    }
  }

}
```

There is a lot going on in this property, but let's take it one piece at a time. First let's look at the `Association` attribute. This attribute has a number of parameters, but these are the basic ones to set. The `IsForeignKey` parameter tells the association that this is a foreign key relationship. The `Storage` parameter describes the name of the class's member that holds the `EntityRef` for this association. The `ThisKey` and `OtherKey` are the columns of the keys on each side of the association. `ThisKey` refers to the name of the column on this class (`Game`) and `OtherKey` refers to the column name on the other side of the association (`Publisher`).

When someone accesses this property, you will return the entity from within the `EntityRef` object as shown in the property getter above.

Finally, the setter has a number of operations. The first and last operations in the setter handle the change management notification just like any column property on your table class. Then it takes the value of the property and sets it to the `Entity` inside the `EntityRef` object. Finally, if the value that is being set is not null, it sets the foreign key ID on the table class so that the column that represents the foreign key is set.

By doing all of this you can have a one-to-many relationship between two table classes. But so far that association is only one-way. In order to complete the association, you may want to have a collection on the `Publisher` table class that represents all the games by that publisher.

Adding the other side of the relationship is similar, but in this case you need an instance of a generic class called `EntitySet`:

```
[Table]
public class Publisher :
  INotifyPropertyChanging, INotifyPropertyChanged
{
  // ...

  EntitySet<Game> _gameSet;
```

```
  [Association(Storage = "_gameSet",
    ThisKey = "Id",
    OtherKey = "_publisherId")]
  public EntitySet<Game> Games
  {
    get { return _gameSet; }
    set
    {
      // Attach any assigned game collection to the collection
      _gameSet.Assign(value);
    }
  }
}
```

The `EntitySet` class wraps around a collection of elements that are associated with a table class. In this case, the `EntitySet` wraps around a collection of games that belong to a publisher. As in the other side of the association, specifying the `Storage`, `ThisKey`, and `OtherKey` helps the context object figure out how the association is created. The only real surprising thing is that when the setter on the `Games` property is called, it attaches whatever games are assigned to it to the set of `Games`. This is normally only called by the context class when executing a query.

Although not obvious, the construction of the `_gameSet` field isn't shown. This needs to be done in the constructor:

```
[Table]
public class Publisher :
  INotifyPropertyChanging, INotifyPropertyChanged
{
  // ...

  public Publisher()
  {
    _gameSet = new EntitySet<Game>(
      new Action<Game>(this.AttachToGame),
      new Action<Game>(this.DetachFromGame));
  }

  void AttachToGame(Game game)
  {
    RaisePropertyChanging("Game");
    game.Publisher = this;
  }

  void DetachFromGame(Game game)
  {
```

```
        RaisePropertyChanging("Game");
        game.Publisher = null;
    }
}
```

In the constructor you must create the `EntitySet`. Note that in the constructor you will also pass in two actions that handle attaching and detaching a game to and from the collection. The purpose of these two actions is to make sure the individual games that are attached/detached also set or clear their association property. In addition, raising the `PropertyChanging` event helps the context object to be very efficient when the association is changing.

Using an Existing Database

Since the underlying database is SQL Server Compact Edition, you may want to use an existing database (.sdf file). To do this you can simply add it to your phone project (as Content), as shown in Figure 8.2.

By marking the database as Content, it will be deployed to the application data folder when your application is installed. Using an existing database means you will have to build your context and class files to match

FIGURE 8.2 SQL Server Compact Edition database as Content

the existing database. Currently there are no tools to build these classes for you.[2]

When you have a database as part of your project, you can refer to it using the appdata moniker when setting up a context object, like so:

```
public class AppContext : DataContext
{
  public AppContext()
    : base("DataSource=appdata:/DB/LocalDB.sdf;File Mode=read only;")
  {
  }

  // ...
}
```

When you use a database directly in the application directory, the database can only be accessed for reading. That means you must include the "file mode" directive in the connection string as well to indicate that the database is read-only.

It is often preferable to use the database in the application directory as a template for your database. In order to do this, you must first copy the database to isolated storage:

```
// Get a Stream of the database from the Application Directory
var dbUri = new Uri("/DB/LocalDB.sdf", UriKind.Relative);
using (var dbStream = Application.GetResourceStream(dbUri).Stream)
{
  // Open a file in isolated storage for writing
  using (var store =
              IsolatedStorageFile.GetUserStoreForApplication())
  using (var file = store.CreateFile("LocalDB.sdf"))
  {

    byte[] buffer = new byte[4096];
    int sizeRead;

    // Write the database out
    while ((sizeRead = dbStream.Read(buffer, 0, buffer.Length)) > 0)
    {
      file.Write(buffer, 0, sizeRead);
    }
  }
}
```

---

2. There are online walkthroughs for using the desktop tools to build the classes and then refactoring them for use on the phone, but it is not a trivial effort.

You can do this by simply copying the database from the application directory using Silverlight's `Application` class to get a stream that contains the database. Then just create a new file in isolated storage (as shown earlier in this chapter) to save the database as a new file. If you copy the database, you can use your context class with the simple isostore moniker to read and write to the newly copied database.

Schema Updates

So, you've created your database-driven application, and now you're ready to update it to a new version. But your users have been dutifully adding data to your database, and you have to change the database. What do you do?

The local database stack for the Windows Phone SDK can help you accomplish this. In the SDK, there is a `DatabaseSchemaUpdater` class that can take an existing database and make additive changes that are safe for the database. These include adding nullable columns, adding tables, adding associations, and adding indexes.

To get started you need to get an instance of the `DatabaseSchema-Updater` class. You retrieve this using the `DataContext` class's `Create-DatabaseSchemaUpdater` method:

```
using (AppContext ctx = new AppContext())
{
  // Grab the DatabaseSchemaUpdater
  var updater = ctx.CreateDatabaseSchemaUpdater();
}
```

This `updater` class allows you to not only make additive changes, but also handle a database version. This gives you a simple way to determine the update level of any database. The `updater` class supports a simple property called `DatabaseSchemaVersion`:

```
var version = updater.DatabaseSchemaVersion;
```

With the database version, you can make incremental updates:

```
// If specific version, then update
if (version == 0)
{
  // Some simple updates (Add stuff, no remove or migrate)
```

```
  updater.AddColumn<Game>("IsPublished");
  updater.DatabaseSchemaVersion = 1;
  updater.Execute();
}
```

The database version always starts at zero and can be changed to a specific database version using the updater. As is typical for a schema change, you would add any of the new columns, tables, indexes, or associations. You would then update the database schema version to ensure that this update can't be executed a second time. Then, over time, you can test for more version blocks. For example, as your application receives more updates, the code might look like this:

```
// If specific version, then update
if (version == 0)
{
  // So simple updates (Add stuff, no remove or migrate)
  updater.AddColumn<Game>("IsPublished");
  updater.DatabaseSchemaVersion = 1;
  updater.Execute();
}
else if (version == 1)
{
  // So simple updates (Add stuff, no remove or migrate)
  updater.AddIndex<Game>("NameIndex");
  updater.DatabaseSchemaVersion = 2;
  updater.Execute();
}
```

You can see that for the first update the version was incremented. Then when the application matured, it added a new update. This is the central use for the database version.

The four different updates that are supported are:

```
updater.AddTable<Genre>();
updater.AddColumn<Game>("IsPublished");
updater.AddIndex<Game>("NameIndex");
updater.AddAssociation<Game>("Genre");
```

When adding a table, the entire table is added (including all columns, associations, and indexes). This means that when you add a table, you do not need to enumerate all the columns, indexes, and associations specifically. Adding a column adds a specific new column. Any new columns must be nullable as there is no way to specify migration to non-nullable columns. Adding an index is based on the name of the index. Finally, an association is added and is based on the property that contains the `Association` attribute.

Complex Schema Changes

If you need to make schema changes that are too complex for the `DatabaseSchemaUpdater` class to accomplish, you will need to do the hard work of creating a new database as well as transferring and migrating the data manually. There is no shortcut for this work.

Database Security

Although you are the only person who will have access to the database that is contained in the application directory or in isolated storage, at times you may want to increase the security of the database by adding levels of security to the database itself.

The two main ways to secure your database are to add a password for access to the database and to enable encryption. When creating a Windows Phone OS 7.1 application, you can do both of these at the same time. You can specify a password for the database directly in the connection string to the database. This is typically specified in the `DataContext` class:

```
public class AppContext : DataContext
{
  public AppContext()
    : base("DataSource=isostore:/Games.sdf;Password=P@ssw0rd!;")
  {
  }

  public Table<Game> Games;

  public Table<Publisher> Publishers;
}
```

When you specify a password before you create the database, the database will be password-protected as well as encrypted. You cannot add a password or encryption after the database has been created. If you decide after your application has been deployed to add a password (and encryption), you will need to create a new database and migrate all your data manually.

Where Are We?

Dealing with data in a smart way when you build a Windows Phone application is important to the success of your application. By understanding the basics of both isolated storage and local database support, you can store the kinds of data you want to store and in the most efficient way possible.

To reiterate, the main benefit of using the database engine is to support querying of the underlying data in an efficient way. If you do not need the query support, you will find that it is generally more efficient to just use isolated storage and serialization to save your data. Making the right decision here can be the difference between a fast, slick application and a slow, groggy mess.

◾ 9 ◾
Multitasking

Phones are different. That's the essential point of this chapter. On these small devices, which we like to think of as small computers, doing too much at one time can hurt the end-user experience. This is why running multiple applications simultaneously is discouraged—in fact, it's disallowed. As you've seen in previous chapters, your application must be able to switch between activated, deactivated, and dormant phases. But what if you need to run tasks in the background? What if you are checking the server for data, downloading files, or performing other tasks that should be done periodically? How can you do that? This chapter will help you figure out how you can accomplish these tasks.

Multitasking

Back in Windows Mobile 6.5 (and before), you had a lot of power as a developer. You could run multiple applications at the same time and, while in the background, actively do work. Although this gave developers a lot of power, it had the side effect of slowing down the phone. Because there was no good monitor of who wrote good or bad code, a single bad application could cause a lot of pain for the phone's user. In fact, most users became familiar with a task management application (or app-killing application) to help minimize this pain.

When Windows Phone was designed, Microsoft decided that this pain was not acceptable; though iPhone using the same strategy certainly enforced the company's reasoning. The first version of Windows Phone did not allow developers to do anything when their applications where not "in the foreground." As you saw in earlier chapters, using tombstoning to mimic the running of multiple applications is a main tenet of the design philosophy around the phone.

Starting in Windows Phone OS 7.1, Microsoft introduced additional functionality to allow background code to run, but your application still cannot run in the background. This functionality includes

- Background agents
- Background Transfer Service

Background agents allow you to run a small piece of code that runs without the benefit of a user interface. Background agents are expected to run in the background and come in several flavors: periodic agents, resource-intensive agents, and audio agents. These agent types allow you to determine whether you want to run a regularly scheduled task, a task that is long-running but only can run in a limited number of scenarios, and a specific type of agent for playing background audio.

The **Background Transfer Service (BTS)** is a set of APIs for asking the phone to download or upload data for your application. While the BTS doesn't let you run arbitrary code, it does handle the common case of sending or receiving data across a network connection.

Background Agents

If you are a desktop or Web developer, you are used to being able to decide when and where your code runs. On the phone you do not have that luxury, so you have to let the operating system dictate exactly when you run your application or some background operations. But how does the operating system handle background operations?

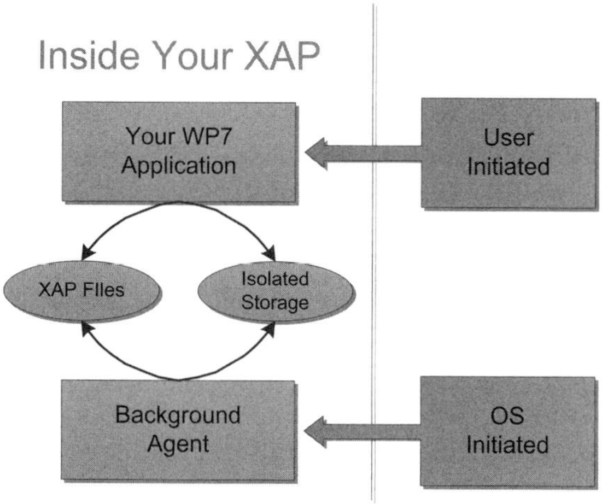

FIGURE **9.1** Relationship between application and scheduled task

Normally when you build a typical phone application, you would submit a .xap file with all the code your application requires to launch the application's user interface. Your application can be launched, but when it is not in the foreground none of your code will execute.

Background agents allow you to supply some code that is executed periodically by the operating system. This code does not have any user interface but shares information with the main application. The information it can share includes isolated storage and application storage (e.g., where the .xap file contents are located), as shown in Figure 9.1.

While the main application is an assembly that contains the startup code, a background agent works similarly. The background agent consists of an additional assembly that is included in the main application's .xap file and contains the code to execute in the background. The main goal of the operating system is to protect the phone from you, the developer. All agents have some very specific limitations, as shown in Table 9.1.

Now that you understand the basic reasons and limitations of using background agents, let's look at each agent type.

TABLE 9.1 Scheduled Task Limitations

Limitation	Description
Forbidden APIs	Scheduled tasks are forbidden from using certain APIs for the phone, including (but not limited to): • `Microsoft.Devices.Camera` class • `Microsoft.Devices.VibrateController` class • `Microsoft.Devices.Radio` namespace • `Microsoft.Devices.Sensors` namespace • Background Transfer Service • `Microsoft.Windows.Controls.WebBrowser` class • `Microsoft.Phone.Tasks` namespace • Scheduled tasks (can't schedule tasks in a scheduled task) • `System.Windows.MessageBox` class • `System.Windows.Clipboard` class • `System.Windows.Controls.MediaElement` class • `System.Windows.Controls.MultiScaleImage` class • `Microsoft.Xna` namespaces • `System.Windows.Navigation` namespace *Note that using the* `GeoCoordinateWatcher` *class is supported but will only supply cached location information. It will not enable A-GPS to determine location.*
Memory	It can only use 5MB of memory total (though audio agents can go as high as 15MB).
Scheduling	It must be rescheduled every two weeks. Usually this means you have to reschedule tasks on every launch of your main application. If your application is not launched every two weeks, the scheduled task will stop being launched.

Periodic Agent

- The **periodic background agent** is a background agent that is meant to execute some code every 30 minutes. To get started you will need a Schedule Task Agent project in your solution. To add a Scheduled Task Agent project, right-click the solution in the Solution Explorer and pick Add | New Project, as shown in Figure 9.2.

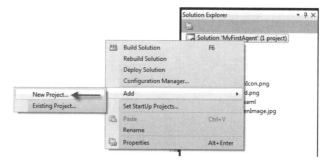

FIGURE 9.2 Adding a new Scheduled Task Agent project

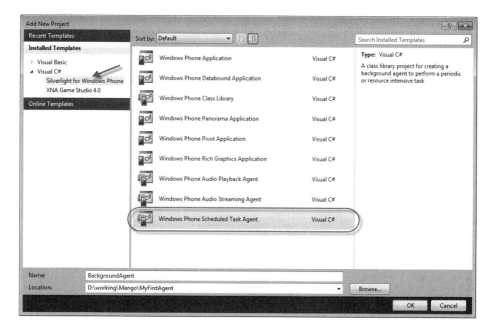

FIGURE 9.3 Picking the Windows Phone Scheduled Task Agent

Once you are in the Add New Project dialog, pick the Windows Phone Scheduled Task Agent project type under the Silverlight for Windows Phone section of your language, as shown in Figure 9.3.

This creates a new project in your solution that contains a single class file:

```
using Microsoft.Phone.Scheduler;

namespace BackgroundAgent
{
```

```
public class ScheduledAgent : ScheduledTaskAgent
{
  protected override void OnInvoke(ScheduledTask task)
  {
    //TODO: Add code to perform your task in background

    NotifyComplete();
  }
}
}
```

The `ScheduledAgent` class that is created overrides a single method (`OnInvoke`) that the operating system calls when the background agent is executed. You should do your work in this method and have the call to `NotifyComplete` be the last line of code in this method. The `Notify-Complete` method tells the operating system that your operation is complete. For example, to save a file with the current time in isolated storage (so your UI app can use it), you could do this:

```
protected override void OnInvoke(ScheduledTask task)
{
  using (var store =
    IsolatedStorageFile.GetUserStoreForApplication())
  using (var file = store.OpenFile("time.txt", FileMode.CreateNew))
  using (var writer = new StreamWriter(file))
  {
    writer.WriteLine(DateTime.Now);
  }

  NotifyComplete();
}
```

Since this code writes directly to isolated storage for the application, that means the main application can read the file we created here. As shown here, you can call the `NotifyComplete` method when you have completed your operation. You may also want to know if the operation fails. You can accomplish this by calling the `Abort` method:

```
try
{
  using (var store =
    IsolatedStorageFile.GetUserStoreForApplication())
  using (var file = store.OpenFile("time.txt", FileMode.CreateNew))
  using (var writer = new StreamWriter(file))
  {
```

```
      writer.WriteLine(DateTime.Now);
    }

    // Successful
    NotifyComplete();
  }
  catch
  {
    // Failure or Aborted
    Abort();
  }
```

You should call either `NotifyComplete` or `Abort` in your agent, to let the runtime (and potentially your application) know whether the task was successfully completed or not.

When you added the new scheduled agent, the project also reached into the main application and added a new section to the WMAppManifest.xml file:

```
<Deployment ...>
  <App ...>
  ...
    <Tasks>
      <DefaultTask Name="_default"
                   NavigationPage="MainPage.xaml" />
      <ExtendedTask Name="BackgroundTask">
        <BackgroundServiceAgent  Specifier="ScheduledTaskAgent"
                                 Name="BackgroundAgent"
                                 Source="BackgroundAgent"
                                 Type="BackgroundAgent.ScheduledAgent"
        />
      </ExtendedTask>
    </Tasks>
    ...
  </App>
</Deployment>
```

Inside the `Tasks` element, the project item added a section called `ExtendedTask`, which is responsible for indicating the project and code for any background tasks. The `ExtendedTask` element is where all agents are registered, including periodic, resource-intensive, and audio agents. Although the `ExtendedTask` is named, the name is not significant. Inside the `ExtendedTask` element is a set of elements that reference the different background agent or agents in your application. Each attribute in the `BackgroundServiceAgent` element has a specific meaning.

- **Name:** This is the name of the element, not referenced in code.
- **Specifier:** This is the type of agent. The different types of specifiers are as follows.
  - *ScheduledTaskAgent:* This is a periodic or resource-intensive task.
  - *AudioPlayerAgent:* This task plays specific songs from a list of audio files.
  - *AudioStreamingAgent*: This task streams audio directly to the phone.
- **Source:** This is the assembly name that contains the background agent.
- **Type:** This is the type of the class that represents the background agent.

This part of the WMAppManifest.xml file is what links your application to the assembly that contains your background task. This means your background agent must be in a separate assembly (as the separate project would indicate). You still have to do a little more work to make your agent actually run in the background.

Before your application can register the background task, you need to make a reference to the new background project. This just requires you to select Add Service Reference and pick the assembly in the Projects tab, as shown in Figure 9.4.

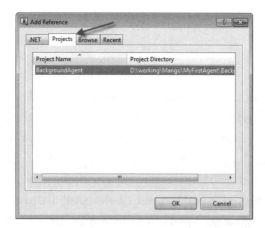

FIGURE 9.4 Adding a reference to the Scheduled Task Agent project

Once your main project has a reference to the background task, you can register it to be executed periodically. To do this you need to create a new instance of your task using the `PeriodicTask` class:

```
// A unique name for your task. It is used to
// locate it in from the service.
var taskName = "MyTask";

// Create the Task
PeriodicTask task = new PeriodicTask(taskName);

// Description is required
task.Description = "This saves some data to Isolated Storage";

// Add it to the service to execute
ScheduledActionService.Add(task);
```

The unique name here is used to locate the service if you need to stop or renew the service, but is not related to the task name in the WMAppManifest.xml file. Once your task is created, you must set the `Description` property as well (at a minimum). After your `PeriodicTask` object is constructed, you can add it to the phone by using the `ScheduledAction-Service`'s `Add` method as shown. This will cause your background task to be periodically executed (every 30 minutes).

Periodic Agent Timing

While the phone attempts to execute your code every 30 minutes, it may execute it several minutes early or late depending on the state of the system (e.g., memory, battery, etc.).

You should set the `Description` property of the `PeriodicTask` class to something significant. The description is an end-user-visible string that is shown in the background task management UI, as shown in Figure 9.5.

Because each periodic task will execute for up to two weeks before it has to be reregistered, you should typically remove and re-create it on every execution of your application:

FIGURE **9.5** The `PeriodicTask`'s description in the management user interface

```
// A unique name for your task. It is used to
// locate it in from the service.
var taskName = "MyTask";

// If the task exists
var oldTask = ScheduledActionService.Find(taskName);
if (oldTask != null)
{
   ScheduledActionService.Remove(taskName);
}

// Create the Task
PeriodicTask task = new PeriodicTask(taskName);

// Description is required
task.Description = "This saves some data to Isolated Storage";

// Add it to the service to execute
ScheduledActionService.Add(task);
```

■ Best Practice

You should not enable any background task by default, and you should always allow your users to disable background tasks in your application.

Now that you have your agent registered, you will need to be able to debug it. The problem on the face of it is that you may not want to wait the 30 minutes for your agent to execute. The `ScheduledActionService` has a way to run the agent immediately so that you can debug it more easily:

```
var taskName = "MyTask";

ScheduledActionService.LaunchForTest(taskName,
  TimeSpan.FromMilliseconds(250));
```

The `LaunchForTest` method takes the name of the task (that you specified earlier when you created the `PeriodicTask`) and a delay before the task is launched. Lastly, since the background task (in this example) was able to write to isolated storage, you can access that data in your main application anytime you want. The background task and your application simply need to communicate by storing information in these shared locations (isolated storage, the Internet, or reading data from the .xap file).

In addition, you might want to alert the user about new information that the background task detected (e.g., a new message is available). You can use the `ShellToast` class to open a toast (or update Live Tiles):

```
protected override void OnInvoke(ScheduledTask task)
{
  // If the Main App is Running, Toast will not show
  ShellToast popupMessage = new ShellToast()
  {
    Title = "My First Agent",
    Content = "Background Task Launched",
    NavigationUri = new Uri("/Views/DeepLink.xaml", UriKind.Relative)
  };
  popupMessage.Show();

  NotifyComplete();
}
```

By using the `ShellToast` class you can alert the user that the background task detected something and give her a chance to launch the application. If the main application is currently running, the `ShellToast` class will not show the pop up and will be reserved to show when your application is not currently being executed.

Creating your own periodic tasks is an easy way to do simple background processing and be able to alert the user to ongoing events and allow her to interact with your application. But there are times when you need a periodic task that consumes more resources. That is where resource-intensive agents come in.

Resource-Intensive Agent

Like a periodic agent, a resource-intensive agent is a repeatable task that performs some discrete process in the background. Unlike the periodic agent, the resource-intensive agent is not meant to be executed very often. In fact, there are a strict set of rules as to when resource-intensive agents are executed. These agents are meant to be able to run for a longer period (up to 10 minutes) and consume more resources (e.g., network and memory). But in order to allow your agent to be executed, the operating system must only allow it when doing so is not detrimental to the phone itself. To that end, the criteria for executing a resource-intensive agent include

- On external power
- When using a noncellular network (e.g., Wi-Fi or plugged into a PC)
- When the minimum battery level is 90%
- When the device must be screen-locked
- When no phone call is active

These rules should imply the fact that resource-intensive agents are meant for syncing large amounts of data or processing that will be accomplished occasionally. These agents will often only be executed once a day, at most. Deciding whether to use an agent (or deciding to use a periodic agent or resource-intensive agent) can increase the overall usefulness of your application.

If the criteria for executing a resource-intensive agent change while the agent is being executed (e.g., a phone call comes in or the phone is removed from external power), the resource-intensive agent will immediately be aborted to allow the user full access to the phone.

> ### ■ Warning
>
> Resource-intensive agents have so many requirements to allow them to be executed that some users will never be able to execute these agents. For example, users who don't dock the phone with a PC or use Wi-Fi will never execute a resource-intensive agent.

Both resource-intensive and periodic agents are scheduled agents. So whether you're adding a periodic or a resource-intensive agent (or even if you need both), you will have a single Scheduled Task Agent project (similar to the one shown previously in the Periodic Agent section).

Registering a resource-intensive agent is virtually identical to registering a periodic agent:

```
// A unique name for your task. It is used to
// locate it in from the service.
var taskName = "IntensiveTask";

// If the task exists
var oldTask = ScheduledActionService.Find(taskName);
if (oldTask != null)
{
  ScheduledActionService.Remove(taskName);
}

// Create the Task
ResourceIntensiveTask task = new ResourceIntensiveTask(taskName);

// Description is required
task.Description = "This does a lot of work.";

// Add it to the service to execute
ScheduledActionService.Add(task);
```

The only real difference is that the class you create is an instance of `ResourceIntensiveTask` (instead of `PeriodicTask`). By specifying that the new task is a resource-intensive task, the operating system will know to only execute the agent when the phone is ready for a resource-intensive operation (e.g., it is in a state as defined by the limitations above).

You can test resource-intensive agents in the same way as well, by calling the `ScheduledActionService`'s `LaunchForTest` method:

```
var taskName = "IntensiveTask";

ScheduledActionService.LaunchForTest(taskName,
  TimeSpan.FromMilliseconds(10));
```

You may want to have both a periodic and a resource-intensive task registered for the phone. The problem is that an application can only have a single background agent (the Scheduled Task Agent project) associated with an application. You can register one (and only one) of each type of scheduled task. This means you can have a periodic task and a resource-intensive task for your application, but since there is only one agent you must discriminate which type of task is being called in your agent by checking the task type, like so:

```
public class ScheduledAgent : ScheduledTaskAgent
{
  protected override void OnInvoke(ScheduledTask task)
  {
    if (task is ResourceIntensiveTask)
    {
      DoHeavyResources();
    }
    else
    {
      DoPeriodic();
    }

    NotifyComplete();
  }

  // ...
}
```

In this way, you can determine the type of task that the OnInvoke is meant to execute. You could discriminate by name as well, but since you can only have one of each type, testing by object type is just as effective.

Audio Agent

Although you can write applications that play audio (as you've seen in prior chapters), you may also want that audio to continue whether your application is running or not. In addition to continuing to play the audio, you may want to integrate with the Universal Volume Control (which allows the user

FIGURE 9.6 The Universal Volume Control (UVC) in action

to move to pause audio as well as move to the previous and next tracks). You can see the Universal Volume Control (UVC) in Figure 9.6.

To be able to have audio play in the background, you need to have an audio agent. There are two types of audio agents: audio playback agents and audio streaming agents. For this example we'll focus on adding an audio playback agent as that is a much more common case.

Background audio works by your application having an audio agent associated with it. This audio agent is solely responsible for handing requests to change the audio in any way (e.g., move tracks, pause, etc.). Your application can have controls that request those changes, but it is ultimately the agent that does the work so that the same agent code is used when the user changes the audio (again, moving tracks, pausing, etc.) through the phone's built-in controls (like the UVC).

To get started you will need to add a new Audio Playback Agent project to your project (like you did for scheduled task agents earlier in this chapter), as shown in Figure 9.7.

FIGURE 9.7 Adding an audio agent to your project

This new project not only creates a project for the audio agent but also modifies your WMAppManifest.xml file (similar to what the Scheduled Task Agent project did earlier):

```
<Deployment ...>
  <App ...>
    ...
    <Tasks>
      <DefaultTask Name="_default"
                   NavigationPage="MainPage.xaml" />
      <ExtendedTask Name="BackgroundTask">
        <BackgroundServiceAgent Specifier="AudioPlayerAgent"
                                Name="MyAudioAgent"
                                Source="MyAudioAgent"
                                Type="MyAudioAgent.AudioPlayer" />
      </ExtendedTask>
    </Tasks>
    ...
  </App>
</Deployment>
```

The BackgroundServiceAgent element added here specifies that the specifier of the agent is an AudioPlayerAgent, which tells the operating system that the `AudioPlayer` class is responsible for playing audio if the application decides to play background audio. Before you can use the new audio agent you will need to add a reference to the audio agent project from your main phone application project, as shown in Figure 9.8.

Unlike the prior agent types, you do not need to register this type of agent with the `ScheduledActionService`. Instead, you will simply use the `BackgroundAudioPlayer` class. This class has a static property called `Instance` that returns the singleton background player object. You can use this class to tell the background audio to start, skip, or stop (or even seek). For example, to implement a button in your application that can play or pause the current audio, you could do the following:

```
private void playButton_Click(object sender, RoutedEventArgs e)
{
  // Tell the agent to play or pause
  if (BackgroundAudioPlayer.Instance.PlayerState !=
      PlayState.Playing)
  {
    BackgroundAudioPlayer.Instance.Play();
  }
  else
  {
    BackgroundAudioPlayer.Instance.Pause();
  }
}
```

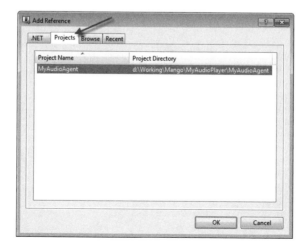

FIGURE 9.8 Making a reference to the audio agent project

What is important to see here is that all your application's code should do is send commands to the background audio player (which is controlled by the agent). When you tell the BackgroundAudioPlayer class to "play," it then routes that command to the agent where all the hard work is done. So, let's look at the skeleton of the audio agent that the new project created for you:

```csharp
public class AudioPlayer : AudioPlayerAgent
{
  protected override void OnPlayStateChanged(
    BackgroundAudioPlayer player,
    AudioTrack track,
    PlayState playState)
  {
    //TODO: Add code to handle play state changes

    NotifyComplete();
  }

  protected override void OnUserAction(
    BackgroundAudioPlayer player,
    AudioTrack track,
    UserAction action,
    object param)
  {
    //TODO: Add code to handle user actions
    // through the application and system-provided UI

    NotifyComplete();
  }

  protected override void OnError(
    BackgroundAudioPlayer player,
    AudioTrack track,
    Exception error,
    bool isFatal)
  {
    //TODO: Add code to handle error conditions

    NotifyComplete();
  }

  protected override void OnCancel()
  {
  }
}
```

The `agent` class contains four different methods that are called in different cases while background audio is playing. Let's start with `OnUserAction`:

```
protected override void OnUserAction(
  BackgroundAudioPlayer player,
  AudioTrack track,
  UserAction action,
  object param)
{
  // May be initiated from the app or the UVC
  switch (action)
  {
    case UserAction.Play:
      {
        PlayCurrentTrack();
        break;
      }
    case UserAction.Pause:
      {
        player.Pause();
        break;
      }
    case UserAction.SkipNext:
      {
        MoveToNextSong();
        break;
      }
    case UserAction.SkipPrevious:
      {
        MoveToPreviousSong();
        break;
      }
  }

  NotifyComplete();
}
```

This method is called whenever the user makes a request to manipulate the currently playing music (including the first request to "play" a song). It passes in the current instance of the `BackgroundAudioPlayer` so that you can manipulate the song to be played directly (as is shown when the `UserAction` is `Pause`). For playing and moving tracks, the list of tracks is owned by the audio agent (not the background audio player). This means you will typically need your audio agent to keep track of the list of tracks and the current position in that list. Any state you need to keep around needs to be static as the specific instance of the audio agent should be stateless (since it can be destroyed or garbage-collected if necessary). So, for this example we need to store both the list of tracks and the track counter as static data:

```
public class AudioPlayer : AudioPlayerAgent
{
    // Song Counter
    static int _currentSong = 0;

    // Songs
    static SongList _songList = new SongList();

    // ...
}
```

This allows you to keep the list of songs that the user has requested (it may be stored in isolated storage or another mechanism for sharing between the audio agent and the application). Then the audio agent can implement the PlayCurrentTrack (which this example calls when a user requests to begin playing), like so:

```
void PlayCurrentTrack()
{
    // Retrieve a Song object that contains the song to play
    var song = _songList.Songs.ElementAt(_currentSong);

    // Build Audio Track
    var track = new AudioTrack(song.Path,
        song.Name,
        song.Artist,
        song.Album,
        song.Art);

    // Instruct the player that this is the current track to play
    BackgroundAudioPlayer.Instance.Track = track;
}
```

This simply returns the current song in the list as a custom type (called Song in this example) and builds a new AudioTrack object from the Song object (by setting a URI to the path to the song as well as descriptive properties for the song name, artist name, album name, and URI to an image for the album art).

Audio Paths

The first parameter to the `AudioTrack` class's constructor is a `Uri` that points at the song. This could be a song in isolated storage (as a relative URI):

```
var path = new Uri("/audio/SomeTrack.mp3", UriKind.Relative);
```

Or the first parameter could be a URI for a song out across a network connection:

```
var path = new Uri("http://shawntwain.com/01-anne.mp3");
```

Only network and isolated storage URIs are supported. If you want to include audio in your .xap file, you'll need to copy it to isolated storage manually.

Once a track has been set the audio will not automatically start playing. Setting the track simply tells the background audio to attempt to find the audio to play. When it finds the audio correctly it will change its play state. Luckily, one of the audio agent's generated methods is called when the play state changes:

```
protected override void OnPlayStateChanged(
  BackgroundAudioPlayer player,
  AudioTrack track,
  PlayState playState)
{
  switch (playState)
  {
    // Track has been selected and is ready to be played
    // (includes starting to download if audio is remote)
    case PlayState.TrackReady:
      {
        player.Play();
        break;
      }
    case PlayState.TrackEnded:
      {
        MoveToNextSong();
        break;
      }
  }
}
```

```
        NotifyComplete();
    }
```

The `PlayState` enumeration has a number of values, but in your case you most want to pay attention to two states.

- **TrackReady:** This indicates that the `BackgroundAudioPlayer` has located and loaded the audio and is ready to actually play.
- **TrackEnded:** This indicates that a track just ended.

When this method is called with `TrackReady`, that's your cue to go ahead and play the audio that has been readied. This is what actually causes the audio to start. You can also handle other states (e.g., `Track-Ended`) to determine what to do next. In this example the code simply moves the track to the next item, eventually calling `PlayCurrentTrack` to load the next track. No matter what happens at this point you should call `NotifyComplete` or `Abort` (like the earlier scheduled task agent example explained).

So far you've seen how to play the first track, but what about changing tracks? Earlier in this section, when the user requested to go to the next or previous track we called methods that would do that work, but what do those methods actually do? They do the simple work of changing the current track number and just telling the audio agent to play the current track once the counter has changed:

```
private void MoveToPreviousSong()
{
    if (--_currentSong < 0)
    {
        _currentSong = _songList.Songs.Count() - 1;
    }

    PlayCurrentTrack();
}

private void MoveToNextSong()
{
    if (++_currentSong >= _songList.Songs.Count())
    {
        _currentSong = 0;
    }
```

```
    PlayCurrentTrack();
  }
```

The work of determining what to set the _currentSong value to is trivial code, once that track number has changed; it calls `PlayCurrentTrack`, which we saw earlier as simply setting the `BackgroundAudioPlayer`'s `Track` property to the new track.

Back in the main application you can use the `BackgroundAudio-Player` class to retrieve the current state of background audio. For example, to show the currently playing song you might check the current `Track`:

```
public partial class MainPage : PhoneApplicationPage
{

  protected override void OnNavigatedTo(NavigationEventArgs e)
  {
    base.OnNavigatedTo(e);

    if (BackgroundAudioPlayer.Instance.Track != null)
    {
      currentSong.Text = BackgroundAudioPlayer.Instance.Track.Title;
    }
  }

  // ...
}
```

When the page is launched it simply tests to see if there is a current track, and if so, it sets the title in the user interface to indicate to the user what the current song is. But since the audio agent is controlling the state of the background audio, you might need to know about the change to the `PlayState` as well. This allows your application to show the current track and enable/disable buttons as necessary:

```
public partial class MainPage : PhoneApplicationPage
{
  // Constructor
  public MainPage()
  {
    InitializeComponent();

    BackgroundAudioPlayer.Instance.PlayStateChanged +=
      Instance_PlayStateChanged;
  }
```

```
void Instance_PlayStateChanged(object sender, EventArgs e)
{
  if (BackgroundAudioPlayer.Instance.Track != null)
  {
    currentSong.Text = BackgroundAudioPlayer.Instance.Track.Title;
  }
}

// ...
}
```

When the event is fired, the play state has changed and (as this example shows) you can change the current audio track if it is valid.

This should give you a general feel for the nature of how audio agents work. The specifics of sharing data between the agent and the application will greatly vary depending on how you want to use background audio, but this way you can allow the user to control the audio without it necessarily being part of the phone's media library.

Background Transfer Service

One of the most common reasons developers want to use background processes for Windows Phone is so that they can download or upload data from the Internet. Having to create an entire agent (and have the user manage that agent) just to accomplish that sort of work seems unnecessary—because it is.

The Windows Phone SDK 7.1 includes a special service called the Background Transfer Service (BTS). This service's job is to allow you to queue up downloads and uploads to be performed by the system. These transfers do not require that your application stay running in order to be performed. Essentially, the service allows you to send it a small number of requests that it will perform on your behalf. But in order to be able to use the service, you need to be aware of a number of requirements and limitations.

Requirements and Limitations

An overarching goal of the BTS is to allow you to perform the work you need to do without impacting the performance of the phone or affecting the user of the phone in a negative way. To meet these goals, the BTS works in very specific ways.

- Transfer protocol:
  - All transfers are made via HTTP or HTTPS.
  - Supported verbs include GET and POST for downloads and POST for uploads.
- Transfer location:
  - All transfers must take place to or from isolated storage in a special subdirectory called /shared/transfers.
  - This directory is created during installation, but if your application deletes it, you must re-create it before any transfers.
  - You can create any files or directories under this subdirectory.
- Forbidden headers:
  - If-Modified-Since
  - If-None-Match
  - If-Range
  - Range
  - Unless-Modified-Since
- Size policies:
  - Maximum upload: 5MB
  - Maximum download via cellular connection: 20MB[1]
  - Maximum download via Wi-Fi on battery: 100MB
  - Maximum download via Wi-Fi on external power: unlimited
- Request limits:
  - Maximum outstanding requests per application: 5
  - Maximum concurrent requests: 2
  - Maximum number of headers per request: 15
  - Maximum size of each HTTP header: 16KB
- Supported networks:
  - 2G, EDGE, Standard GPRS: not supported
  - 3G: supported but must have minimum 50Kbps throughput
  - Wi-Fi/PC connection: supported but must have minimum 100Kbps throughput

---

1. If a download exceeds 20MB, the remaining data will require download via Wi-Fi or PC connection and external power.

Requesting Transfers

To get the BTS to perform a transfer you must first create a new request. A request takes the form of a `BackgroundTransferRequest` object. At a minimum you must specify the source and destination of your request like so:

```
// Determine the source and destination
var serverUri = new Uri("http://shawntwain.com/01-Anne.mp3");
var isoStoreUri = new Uri("/shared/transfers/01-Anne.mp3",
                          UriKind.Relative);

// Create the Request
var request = new BackgroundTransferRequest(isoStoreUri, serverUri);
```

The constructor for the request can accept two URIs that specify where to copy from and where to copy to. Typically this takes the form of an Internet URI for the source (for downloading) and a relative URI to the special /shared/transfers directory in isolated storage. In order to make the request you can simply add this to the requests on the service (e.g., the `BackgroundTransferService` class):

```
BackgroundTransferService.Add(request);
```

This will queue up the request and have the service perform the transfer for you. To request an upload the method is similar, but the source and destination are reversed as well as specifying the method:

```
// Determine the source and destination
var serverUri = new Uri("http://shawntwain.com/01-Anne.mp3");
var isoStoreUri = new Uri("/shared/transfers/01-Anne.mp3",
                          UriKind.Relative);

// Create the Request
var request = new BackgroundTransferRequest(serverUri);

// Set options
request.UploadLocation = isoStoreUri;
request.Method = "POST";

// Queue the Request
BackgroundTransferService.Add(request);
```

The constructor for the `BackgroundTransferRequest` that takes two arguments is specifically for downloading files from the server. When you want to request an upload, you need to create the request by specifying the server URI that represents where the upload is going. Then you need to specify the UploadLocation as a URI that represents the path to files in the special transfer directory from which to upload.

In addition to simply specifying the source and destination for your request, you can also specify other optional properties on the BackgroundTransferRequest class, including the following.

- **Headers:** This allows you to set specific headers for the HTTP transfer.
- **Tag:** This allows you to set an arbitrary string that contains additional information that you may use when retrieving the request.
- **TransferPreferences:** This is an enumeration that specifies what circumstances are allowed.
  - *None:* Only allow transfers while on external power and using a high-speed network (e.g., Wi-Fi or PC connection). This is the default.
  - *AllowCellular:* Allow the request while on a cellular network but must be on external power.
  - *AllowBattery:* Allow the request on Wi-Fi but do not require an external power source.
  - *AllowCellularAndBattery:* Allow the transfer on cellular and while on battery. You should only use this for small and immediately needed requests. Use judiciously!

Requesting the transfers is adequate, but it is also up to your application to monitor the transfers (and possibly let the user know the status of the transfer). That is where monitoring of your own transfers becomes necessary.

Monitoring Requests

The BackgroundTransferService class supports a static property called Requests that represents an enumerable list of all the current requests (including those that have completed and those that have failed). Because access to the requests is through a property, you may be lured into expecting that the collection represents the overall status of requests, but that is not how it works. When you access the property it returns a copy of the current state of the requests, but these are not updated as the requests are processed. So even though you can use data binding to show the user the list of requests and their current status, they will not update until you retrieve the list of results again.

Retrieving the Requests

Every time you access the `Requests` property on the `Background-TransferService` it makes a copy of the current state of each request. This process is not cheap, so it is recommended that you do not access this property in a tight loop or based on a subsecond timer.

Let's look at a strategy for keeping the user apprised of the status of requests. To start, you should keep a local cache of the last requests you retrieved from the `BackgroundTransferService` class, like so:

```
public partial class MainPage : PhoneApplicationPage
{
  // Constructor
  public MainPage()
  {
    InitializeComponent();
  }

  IEnumerable<BackgroundTransferRequest> _requestCache = null;

  // ...
}
```

You should retrieve the requests when someone navigates to the page, so you can show the status to the user:

```
protected override void OnNavigatedTo(NavigationEventArgs e)
{
  base.OnNavigatedTo(e);

  // Update the Page
  RefreshBindings();
}
```

In this example, the `RefreshBindings` method is used to retrieve the requests again, as well as bind them to the user interface:

```
void RefreshBindings()
{
  // Dispose of the old cache to stop memory leaks
  if (_requestCache != null)
  {
    foreach (var request in _requestCache)
    {
      // Clean up the cache to prevent leaks
      request.TransferProgressChanged -=
        request_TransferProgressChanged;
      request.TransferStatusChanged -= request_TransferStatusChanged;
      request.Dispose();
    }
  }

  // Get the Updated Requests
  _requestCache = BackgroundTransferService.Requests;

  // Wire up the events
  foreach (var request in _requestCache)
  {
    request.TransferProgressChanged +=
      request_TransferProgressChanged;
    request.TransferStatusChanged += request_TransferStatusChanged;
  }

  // Rebind
  filesListBox.ItemsSource = _requestCache;
}
```

The first part of this method takes any existing cache (as this will be called in a number of cases as the transfers change in status) and cleans them so that they don't leak resources. The returned requests support the `IDisposable` interface, so you must call `Dispose` on each of them. In addition, this example is unwiring two events that we are wiring up every time we retrieve the requests. Once the cleanup is complete, the code retrieves the current state of the requests. When it has the new requests it wires up two events, one that indicates that the request's progress has changed (e.g., the transfer is proceeding) and one that is fired when the status of the transfer changes (succeeded or failed). Lastly, the new cache is rebound to the user interface (a `ListBox` in this case).

On the face of it, this is a lot of work where you might be used to just binding to a collection that simply changes the underlying data using binding interfaces (e.g., `INotifyPropertyChanged` and `INotifyCollectionChanged`). This is not how the `BackgroundTransferService`'s requests work. You must rebind them on every update.

To ensure that the page is updated as the progress is changed, the handler for each request's `TransferProgressChanged` event simply calls the `RefreshBinding` method:

```
void request_TransferProgressChanged(object sender,
                                     BackgroundTransferEventArgs e)
{
  // Update the Page
  RefreshBindings();
}
```

This does mean that the underlying data is being refreshed quite a lot, but that's necessary to update the user with the latest progress information. You can decide to only update it as transfers complete (which happens less often). This example handles that event as well:

```
void request_TransferStatusChanged(object sender,
                                   BackgroundTransferEventArgs e)
{
  // Limited to 5 requests
  // So remove it when it's done
  if (e.Request.TransferStatus == TransferStatus.Completed)
  {
    if (BackgroundTransferService.Find(e.Request.RequestId) != null)
    {
      BackgroundTransferService.Remove(e.Request);
    }
    e.Request.Dispose();
  }

  // Update the Page
  RefreshBindings();
}
```

The difference here is that not only is the code calling `Refresh-Bindings`, but if a request is complete it is being removed from the service. While removing the completed items is suggested, you can decide exactly when to do this. The reason that completed requests are removed automatically is so that when the user relaunches your application, you can see what requests succeeded; therefore (as this example shows), you will need to remove those requests. This becomes important because (as stated earlier) you can only have five requests at a time. If you have five requests in the service (even if they are all complete), the sixth request will throw an exception. It is up to you to manage the list of transfers.

When displaying the status of the requests, the `BackgroundTransfer-Request` objects that are returned include several pieces of read-only data.

- **RequestId:** This is a generated GUID that can be used to track the request.
- **TotalBytesToReceive:** This is the total number of bytes to be downloaded in the request. This value is `0` until the request has started.
- **BytesReceived:** This is the number of bytes downloaded so far.
- **TotalBytesToSend:** This is the total number of bytes to be uploaded in the request.
- **BytesSent:** This is the number of bytes sent so far.
- **TransferStatus:** This is an enumeration that indicates the current state of the transfer.
  - *None:* The system has not queued the request yet.
  - *Transferring:* The request is being performed.
  - *Waiting:* Typically this means it is waiting for other pending transfers to be completed.
  - *WaitingForWiFi:* The request is queued, but it cannot start until a Wi-Fi connection is available.
  - *WaitingForExternalPower:* The request is queued, but it is waiting for external power to be available (e.g., the phone is on battery).
  - *WaitingForExternalPowerDueToBatterySaverMode:* The battery is low and all requests have been suspended.
  - *WaitingForNonVoiceBlockingNetwork:* The request is queued but it is waiting for a higher-speed cellular network or Wi-Fi to be available to complete the request.
  - *Paused:* The BTS has stopped the request temporarily.

- *Completed:* The request is complete. You should check the `TransferError` to ensure that the request was successful.
- *Unknown:* The system is unable to determine the state of the transfer.

- **StatusCode:** This is the HTTP status code (a number) that indicates what was returned by the server.
- **TransferError:** This is the exception (if any) that was encountered during the transfer. You should check to see that the `TransferError` is null when a transfer is complete to ensure that the transfer completed successfully.

The way your application manages requests is crucial if you are going to use the BTS, as your application is the only place to manage your requests. The user does not have an operating system management screen for the BTS.

While keeping the user apprised of the status of the transfers is important (no matter how you let her know about the status), the work is worth the effort as building an efficient and robust download and upload system expressly for your application is not an easy task. By relying on the BTS to accomplish these types of transfers, it should be easier to write your application.

Where Are We?

Windows Phone generally wants to dissuade you from affecting the performance of the overall phone experience. The multitasking support on the phone gives you a number of performance- and resource-friendly ways to accomplish much of what you might need to accomplish by running background processes. If you're coming from desktop or server development, at first the challenges posed by these limitations can be daunting. But you should be able to achieve most of the types of applications you want to build without hurting the overall experience on the phone. By using background agents and the BTS, most of your needs should be met.

10

Services

Windows Phone is a small, mostly connected device, meaning that most of the time you can expect to be able to access the Internet from the phone. In many ways, Internet access is what has really changed smartphones over the past few years. When you are building your Windows Phone application, you can use the Internet to get information, assets, or whatever your application requires. In this chapter we will review the different ways in which you can use services, from the generic use of REST or Web services, to specific examples of using push notifications and Windows Live services.

> **■ Note**
>
> The free tools included in the Windows Phone SDK 7.1 are adequate for most of the development story on the phone. But if you need to write your own services to be consumed by the phone, there is no way to get Visual Studio Express to create those Web projects for you. There are several approaches to make this work, but in this chapter we will use Visual Studio Professional to create the services to simplify the code.

The Network Stack

The network stack in Silverlight (and on the phone) allows you to access network (e.g., Internet) resources. Silverlight dictates that all network calls must be performed asynchronously. This is to ensure that an errant network call does not tie up the user interface (or more important, make the phone think your application has locked up or crashed). If you are new to asynchronous programming, this will be a good place to start learning.

The `WebClient` Class

In most cases the Silverlight framework helps out with classes that follow a common pattern that includes a method that ends with the word *Async* and starts an asynchronous execution, and an event that is called when the execution is complete.

One class that follows this pattern is the `WebClient` class. As you can see, it is a simple way to make a basic Web request in Silverlight:

```
public partial class MainPage : PhoneApplicationPage
{
  // Constructor
  public MainPage()
  {
    InitializeComponent();

    // Create the client
    WebClient client = new WebClient();

    // Handle the event
    client.DownloadStringCompleted += client_DownloadStringCompleted;

    // Start the execution
    client.DownloadStringAsync(new Uri("http://wildermuth.com/rss"));
  }
}
```

In this example the `WebClient` class is created, the `Download-StringCompleted` event is wired up, and the `DownloadStringAsync` method is called to start the download. When the download is complete, the event is fired like so:

```
void client_DownloadStringCompleted(object sender,
                        DownloadStringCompletedEventArgs e)
```

```
{
  // Make sure the process completed successfully
  if (e.Error == null)
  {
    // Use the result
    string rssFeed = e.Result;
  }
}
```

In the event argument, you are passed any exception that is thrown when the download is being attempted. This is returned to you as the `Error` property. In the preceding code, the error is `null`, which means no error occurred and the `Result` property will be filled with the result of the network call. This pattern means you have to get used to the process of performing these types of operations asynchronously. While still asynchronous, you can simplify the pattern a little by using a lambda for the event handler:

```
// Create the client
WebClient client = new WebClient();

// Handle the event
client.DownloadStringCompleted += (s, e) =>
  {
    // Make sure the process completed successfully
    if (e.Error == null)
    {
      // Use the result
      string rssFeed = e.Result;
    }
  };

// Start the execution
client.DownloadStringAsync(new Uri("http://wildermuth.com/rss"));
```

In this case, we're writing the handling of the completed event as an inline block of code, which makes the process feel more linear. If you're not familiar with lambda expressions, please see the Microsoft documentation[1] for more details.

The `WebClient` class is the starting point for most simple network calls. The class matches up an "Async" method and a "Completed" event for several types of operations, including the following.

---

1. http://msdn.microsoft.com/en-us/library/bb397687.aspx

- **DownloadString:** This downloads a text result and returns a string.
- **OpenRead:** This downloads a binary stream and returns a `Stream` object.
- **UploadString:** This writes text to a server.
- **OpenWrite:** This writes a binary stream to a server.

The code for using these other types of calls looks surprisingly similar. For example, to download a binary stream (e.g., any nontext object, such as an image), use this code:

```
// Create the client
WebClient client = new WebClient();

// Handle the event
client.OpenReadCompleted += (s, e) =>
  {
    // Make sure the process completed successfully
    if (e.Error == null)
    {
      // Use the result
      Stream image = e.Result;
    }
  };

// Start the execution
client.OpenReadAsync(new

Uri("http://wildermuth.com/images/headshot.jpg"));
```

Although the pattern is the same, the event and method names have changed and the result is now a stream instead of a string.

■ HttpWebRequest/HttpWebResponse

If you've written .NET networking code before, you may be used to working with the `HttpWebRequest` and `HttpWebResponse` classes. Though since the framework is Silverlight, these classes only support asynchronous execution. These classes are supported, but the `WebClient` class is more commonly used because it always fires its events on the same thread as they were originally called (usually the UI thread).

Accessing Network Information

Before you can execute network calls you must have access to the network. On the phone you can test for whether connectivity is supported as well as the type of connectivity (which may help you decide how much data you can reliably download onto the phone). Although the .NET Framework contains several APIs for accessing network information, there is a specialized class in the Windows Phone SDK 7.1 that supplies much of this information all in one place. The `DeviceNetworkInformation` class allows you to access this network information.

The `DeviceNetworkInformation` class supports a number of static properties that will give you information about the phone's network.

- **IsNetworkAvailable:** This is a Boolean value that indicates whether any network is currently available.

- **IsCellularDataEnabled:** This is a Boolean value that indicates whether the phone has enabled cellular data (as opposed to Wi-Fi data).

- **IsCellularDataRoamingEnabled:** This is a Boolean value that indicates whether the phone has enabled data roaming.

- **IsWifiEnabled:** This is a Boolean value that indicates whether the phone has enabled Wi-Fi on the device.

- **CellularMobileOperator:** This returns a string that contains the name of the mobile operator.

You can use this class to determine whether an active network connection is available:

```
if (DeviceNetworkInformation.IsNetworkAvailable)
{
  status.Text = "Network Found";
}
else
{
  status.Text = "Network not Found";
}
```

You can also access an event that indicates that the network information has changed:

```
public partial class MainPage : PhoneApplicationPage
{
  // Constructor
  public MainPage()
  {
    InitializeComponent();

    DeviceNetworkInformation.NetworkAvailabilityChanged +=
      DeviceNetworkInformation_NetworkAvailabilityChanged;
  }

  void DeviceNetworkInformation_NetworkAvailabilityChanged(
    object sender, NetworkNotificationEventArgs e)
  {
    switch (e.NotificationType)
    {
      case NetworkNotificationType.InterfaceConnected:
        status.Text = "Network Available";
        break;
      case NetworkNotificationType.InterfaceDisconnected:
        status.Text = "Network Not Available";
        break;
      case NetworkNotificationType.CharacteristicUpdate:
        status.Text = "Network Configuration Changed";
        break;
    }
  }
  // ...
}
```

The DeviceNetworkInformation class's NetworkAvailability-
Changed event fires whenever the network changes. You can see in this
example that you can access the NotificationType from the Network-
NotificationEventArgs class to see whether a network connection
was just connected, disconnected, or just changed its configuration. In
addition, this event passes in the network type:

```
public partial class MainPage : PhoneApplicationPage
{
  // Constructor
  public MainPage()
  {
    InitializeComponent();

    DeviceNetworkInformation.NetworkAvailabilityChanged +=
      DeviceNetworkInformation_NetworkAvailabilityChanged;
  }
```

```
void DeviceNetworkInformation_NetworkAvailabilityChanged(
  object sender, NetworkNotificationEventArgs e)
{
  switch (e.NetworkInterface.InterfaceSubtype)
  {
    case NetworkInterfaceSubType.Cellular_1XRTT:
    case NetworkInterfaceSubType.Cellular_EDGE:
    case NetworkInterfaceSubType.Cellular_GPRS:
      status.Text = "Cellular (2.5G)";
      break;
    case NetworkInterfaceSubType.Cellular_3G:
    case NetworkInterfaceSubType.Cellular_EVDO:
    case NetworkInterfaceSubType.Cellular_EVDV:
      status.Text = "Cellular (3G)";
      break;
    case NetworkInterfaceSubType.Cellular_HSPA:
      status.Text = "Cellular (3.5G)";
      break;
    case NetworkInterfaceSubType.WiFi:
      status.Text = "WiFi";
      break;
    case NetworkInterfaceSubType.Desktop_PassThru:
      status.Text = "Desktop Connection";
      break;
    }
  }
  // ...
}
```

By using the NetworkInterfaceInfo class's NetworkInterface-SubType enumeration, you can determine the exact type of network connection on the phone. The NetworkInterfaceInfo class also gives you access to several other useful properties, including the following.

- **Bandwidth:** This is an integer that specifies the speed of the network interface.
- **Characteristics:** This is an enumeration that specifies whether the phone is currently roaming or not.
- **Description:** This is a description of the network interface.
- **InterfaceName:** This is the name of the network interface.
- **InterfaceState:** This states whether the network interface is connected or disconnected.

> ### ■ Ethernet Network Connections
>
> Ethernet connections are only available if the phone is hooked up to a computer via a USB cable. This is a common way to tell if the user is plugged into a computer.

By using this `DeviceNetworkInformation` class, you can have access to much of the network information that you will need for your application.

Consuming JSON

Many types of services on the Internet are easy to consume with JavaScript on Web pages and use a special type of data called JSON. JSON stands for JavaScript Object Notation. This format is useful as it is very easy to consume from JavaScript and tends to be smaller than XML for certain types of data. As you work with different Internet services you will find a number of APIs that support JSON. For more information on how JSON is structured, visit http://json.org.

Consuming JSON on the phone is fairly straightforward. You have two approaches.

- Serialize JSON to and from managed objects (e.g., classes).
- Parse JSON much like you would XML.

Before you can handle the JSON, you have to retrieve it. The network stack can do this for you. For example, if you wanted to retrieve statuses from Twitter's live feed, you could make a call using the `WebClient` class to retrieve the information:

```
// Create the client
WebClient client = new WebClient();

// Handle the event
client.DownloadStringCompleted += (s, e) =>
  {
    // Make sure the process completed successfully
    if (e.Error == null)
```

```
    {
      // Retrieve the JSON
      string jsonString = e.Result;


    }
  };

  // Start the execution
  string api = "http://api.twitter.com/1/statuses/public_timeline.json";
  client.DownloadStringAsync(new Uri(api));
```

Retrieving data from services that expose their data via JSON is like any other type of network request; therefore, you can use the `WebClient` class to accomplish this. As this example shows, you can use the `Download-String` API to retrieve JSON data.

While the phone does have built-in support for serializing objects to and from JSON, the current version is not very forgiving. The phone does not have built-in support for parsing JSON directly. In both of these cases, I suggest you look at using an open source library that has existed for .NET for quite a while and supports Windows Phone: Json.NET. You can start by downloading the Json.NET libraries from http://json.codeplex.com. The following sections show you how to both serialize objects to and from JSON as well as parse JSON using the Json.NET library.

Using JSON Serialization

The Json.NET libraries support serialization through a class called `Json-Convert` which allows you to serialize and deserialize objects to and from JSON. Passing in an object to `JsonConvert`'s `SerializeObject` method will return a string made up of JSON:

```
var guy = new Person()
{
  FirstName = "Shawn",
  LastName = "Wildermuth",
  BirthDate = new DateTime(1969, 4, 24)
};

string json = JsonConvert.SerializeObject(guy);

// Returns:
// {
//    "FirstName":"Shawn",
```

```
//    "LastName":"Wildermuth",
//    "BirthDate":"\/Date(-21758400000-0400)\/"
// }
```

The `SerializeObject` method takes a single object but will serialize an entire object tree or collection. The passed-in object is the start of the serialization. To reverse the process, you would use the generic `Deserialize-Object` method:

```
string theJson = @"{
  ""FirstName"":""Shawn"",
  ""LastName"":""Wildermuth"",
  ""BirthDate"":""\/Date(-21758400000-0400)\/""
}";

Person recreated = JsonCon
vert.DeserializeObject<Person>(theJson);
```

Put this together with the `WebClient` class and you can interact with JSON-powered services (usually REST services). For example, to call Twitter to get the latest tweets in the public feed, you can call a REST interface like so:

```
// Create the client
WebClient client = new WebClient();

// Handle the event
client.DownloadStringCompleted += (s, e) =>
{
  // Make sure the process completed successfully
  if (e.Error == null)
  {
    // Use the result
    string jsonString = e.Result;

    Tweet[] tweets =
      JsonConvert.DeserializeObject<Tweet[]>(jsonString);

    tweetList.ItemsSource = tweets;

  }
};

// Start the execution
string req = "http://api.twitter.com/1/statuses/public_timeline.json";
client.DownloadStringAsync(new Uri(req));
```

By using the `WebClient` class's networking to request data from Twitter, the JSON returned can be turned into objects directly. When you use Json.NET's `DeserializeObject` method, it is smart enough to try to match the fields of the managed object to the JSON properties. It does not require that the entire object graph in JSON be modeled as classes. This means that when it maps the JSON object to your class, it will only include properties that both your class and the JSON object have in common.

Parsing JSON

Another option with dealing with JSON is to parse it and consume it without having to create classes to hold the data. The Json.NET library includes a couple of classes that simplify this: `JObject` and `JArray`. For example, to read a simple object, just call the `JObject` class's `Parse` method:

```
string theJson = @"{
  ""FirstName"":""Shawn"",
  ""LastName"":""Wildermuth"",
  ""BirthDate"":""\/Date(-21758400000-0400)\/""
}";

JObject jsonObject = JObject.Parse(theJson);

string firstName = jsonObject["FirstName"].Value<string>();
string lastName = jsonObject["LastName"].Value<string>();
DateTime birthDate = jsonObject["BirthDate"].Value<DateTime>();
```

Parsing the JSON into an instance of `JObject` gives you first-class access to the properties of the JSON. You can retrieve the specific values in the JSON by using the `Value` method. The generic version of the `Value` method lets you specify the type of data you expect to retrieve, as shown above.

Although this example shows you a simple object, you can retrieve collections in JSON as well. The `JArray` class is how you parse and consume JSON collections:

```
string jsonCollection = @"[
{
  ""FirstName"":""Shawn"",
  ""LastName"":""Wildermuth"",
  ""BirthDate"":""\/Date(-21758400000-0400)\/""
},
```

```
{
  ""FirstName"":""Pennie"",
  ""LastName"":""Wildermuth"",
  ""BirthDate"":""\/Date(-21758400000-0400)\/""
},
]";

JArray jsonArray = JArray.Parse(jsonCollection);

foreach (JObject item in jsonArray)
{
  string firstName = item["FirstName"].Value<string>();
  string lastName = item["LastName"].Value<string>();
  DateTime birthDate = item["BirthDate"].Value<DateTime>();
}
```

Since the JSON being parsed here is an array of items (note the square bracket that starts the JSON) the JArray class will parse this into an array of JObject items, as shown here. This also means that JSON objects that contain collections will allow you to iterate through them as JArray objects, like so:

```
string json = @"{
  ""FirstName"":""Shawn"",
  ""LastName"":""Wildermuth"",
  ""BirthDate"":""\/Date(-21758400000-0400)\/"",
  ""FavoriteGames"" :
  [
    {
      ""Name"" : ""Halo 3"",
      ""Genre"" : ""Shooter""
    },
    {
      ""Name"" : ""Forza 2"",
      ""Genre"" : ""Racing""
    },
  ]
}";

JObject guy = JObject.Parse(json);

string firstName = guy["FirstName"].Value<string>();
string lastName = guy["LastName"].Value<string>();
DateTime birthDate = guy["BirthDate"].Value<DateTime>();

foreach (JObject item in guy["FavoriteGames"])
{
```

```
    string name = item["Name"].Value<string>();
    string genre = item["Genre"].Value<string>();
}
```

While the object being parsed is a `JObject`, the collection in the JSON (accessed via the indexer, like any other field) would yield a `JArray` object (that can be iterated through). In this way, you can go through a complex object graph in JSON as needed.

The Json.NET library even lets you execute LINQ queries against parsed JSON. You can do this by writing LINQ queries over `JArray` objects like so:

```
string json = @"[
{
  ""FirstName"":""Shawn"",
  ""LastName"":""Wildermuth"",
  ""BirthDate"":""\/Date(-21758400000-0400)\/"",
  ""FavoriteGames"" :
    [
      {
        ""Name"" : ""Halo 3"",
        ""Genre"" : ""Shooter""
      },
      {
        ""Name"" : ""Forza 2"",
        ""Genre"" : ""Racing""
      },
    ]
},
{
  ""FirstName"":""Pennie"",
  ""LastName"":""Wildermuth"",
  ""BirthDate"":""\/Date(-21758400000-0400)\/"",
  ""FavoriteGames"" : []
},
]";

JArray people = JArray.Parse(json);

var qry = from p in people
          where p["FavoriteGames"].Count() > 0
          select p;

var peopleWithGames = qry.ToArray();
```

In this example the query is searching for people who have at least one item in their `FavoriteGames` field in the JSON. This returns the raw

Json.NET objects so that you can continue to pull out the data manually as shown in earlier examples, though you can use LINQ to project into classes as well. This technique is commonly used when you want to query the JSON result but end up with a subset of the JSON results as managed objects.

The major reason you want to end up with managed objects (e.g., instances of classes) is to use data binding. Although data binding can work with `JObject` objects, there is no way to have the name in a binding element do a lookup into the field of a JSON object. In most cases you will want to project these into managed objects like so:

```
// Create the client
WebClient client = new WebClient();

// Handle the event
client.DownloadStringCompleted += (s, e) =>
{
  // Make sure the process completed successfully
  if (e.Error == null)
  {
    JArray tweets = JArray.Parse(jsonString);

    var qry = from t in tweets
              where t["source"].Value<string>() == "web"
              let username = t["user"]["screen_name"].Value<string>()
              orderby username
              select new Tweet()
              {
                Text = t["text"].Value<string>(),
                User = new User()
                {
                  Screen_Name = username,
                }
              };

    tweetList.ItemsSource = qry.ToList();

  }
};

// Start the execution
string req = "http://api.twitter.com/1/statuses/public_timeline.json";
client.DownloadStringAsync(new Uri(req));
```

By projecting into a class structure, you can end up with classes that easily handle data binding directly to XAML objects. This approach is often better than using serialization as you may only need a subset of the result of the data that you retrieved from a REST-based interface. This way, you can use LINQ to filter, sort, and shape your data into objects that better represent what you want to use on the phone.

Web Services

When writing applications for Windows Phone you can use existing or new Web services to communicate with the server. Visual Studio allows you to make references to existing Web services, though the free version of Visual Studio for the phone (Visual Studio Express for Windows Phone) does not support a method of creating new Web projects to host services. If you need to create your own services to be hosted on your own servers you will need Visual Studio Professional or better.

You can consume a Web service by adding a service reference to your project. In Visual Studio, you can right-click the phone project and select Add Service Reference, as shown in Figure 10.1.

This will open a dialog where you can enter a service's address or just discover your own Web services. The dialog has an address bar in which you can simply enter the address of the Web service; when you click the Go button, Visual Studio will find your service, as shown in Figure 10.2. Once your service is discovered, you can specify the namespace and click OK to

Figure 10.1 Adding a service reference

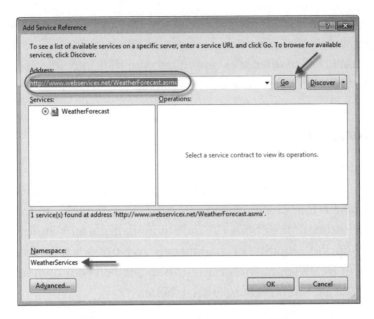

generate a set of classes that will let you call the Web service. The address in this example is a service that returns the weather based upon zip code.[2]

This will add the code that is required to interact with the Web service.

■ Source Code for Service References

In the project tree, a new node called Service References will show every service you've added a reference to. Normally these are single nodes for each service, but if you want to look at the code that is generated you can click the Show All Files button in the Solution Explorer to show all the generated files. The file with the code in it is called Reference.cs, as shown in Figure 10.3.

After the service reference is added, you will have a number of new classes and interfaces that are generated in the namespace that you specified in the dialog. The most important of these interfaces is a WebClient-like class that exposes all the methods of the Web service as asynchronous

---

2. www.webservicex.net/WeatherForecast.asmx

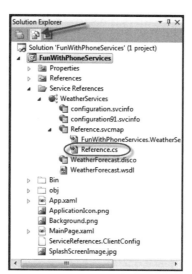

FIGURE 10.3 Service files displayed

methods. The name of this class depends on how the service was written, but it always ends in "Client". In this example the class is called `WeatherForecastSoapClient`. The service contains an operation called `GetWeatherByZipCode`, which the service reference splits into a `Completed` event and an `Asynchronous` call, as shown below:

```
// Open with default address/binding information
var client = new WeatherForecastSoapClient();

// Handle the Completed Event
client.GetWeatherByZipCodeCompleted += (s, a) =>
  {
    if (a.Error == null)
    {
      theList.ItemsSource = a.Result.Details;
    }
    else
    {
      MessageBox.Show("Failed to get weather data.");
    }
  };

// Get Weather Asynchronously
client.GetWeatherByZipCodeAsync("30307");
```

Calling a Web service follows the same pattern as the `WebClient` class shown earlier in this chapter. The results of the Web service are passed into the `Completed` event as the `Result` property of the second argument in the event handler. In this case the result that the Web service returns contains a list of details for each day. By assigning this to the `ItemsSource` of a control in the XAML, data binding will show the list of details.

In the example above, you may have noticed that you did not have to specify the server address. When the service reference was added, a new file was added to your project, called `ServiceReferences.Client-Config`. This file is the configuration for your service:

```
<configuration>
  <system.serviceModel>
    <bindings>
      <basicHttpBinding>
        <binding name="WeatherForecastSoap"
                 maxBufferSize="2147483647"
                 maxReceivedMessageSize="2147483647">
          <security mode="None" />
        </binding>
      </basicHttpBinding>
    </bindings>
    <client>
      <endpoint
         address="http://www.webservicex.net/WeatherForecast.asmx"
         binding="basicHttpBinding"
         bindingConfiguration="WeatherForecastSoap"
         contract="WeatherServices.WeatherForecastSoap"
         name="WeatherForecastSoap" />
    </client>
  </system.serviceModel>
</configuration>
```

The key part of this configuration for the phone is the address of the endpoint (shown in bold). For a public Web service such as the one in the example, you don't need to change this. But for services that you are going to host yourself, you will likely have a test address (probably on your own machine) and need to change this when you deploy your application to specify a production machine. The best solution is to create a duplicate endpoint section and name the endpoints something significant:

```
<client>
  <endpoint address="http://www.webservicex.net/WeatherForecast.asmx"
            binding="basicHttpBinding"
```

```
                bindingConfiguration="WeatherForecastSoap"
                contract="WeatherServices.WeatherForecastSoap"
                name="Production" />
      <endpoint address="http://localhost:8888/WeatherForecast.asmx"
                binding="basicHttpBinding"
                bindingConfiguration="WeatherForecastSoap"
                contract="WeatherServices.WeatherForecastSoap"
                name="Debugging" />
    </client>
```

When you create the client object, you can specify the name of the endpoint. For example, if you wanted to use the production Web server in release builds you could do the following:

```
        // Open with default address/binding information
#if DEBUG
        var client = new WeatherForecastSoapClient("Debugging");
#else
        var client = new WeatherForecastSoapClient("Production");
#endif
```

> ### ■ Creating Your Own Services
>
> Although you may find that using Web services across the Internet is a common approach, at some point you will need to write your own way of communicating with the phone. Unfortunately, you can't do this with the Visual Studio 2010 Express for Windows Phone that comes free with the phone tools. You could use Visual Web Developer 2010, but that would require that you run and coordinate two developer tools to get it to work. This is awkward and difficult. In general, it is better to get Visual Studio 2010 Professional (or better) so that you can create your own Web projects that can host your own Web services in the same solution as your phone applications.

Consuming OData

Open Data Protocol[3] (OData) is a standard way of exposing relational data across a service layer. OData is built on top of other standards including HTTP, JSON, and Atom Publishing Protocol (AtomPub, which is an XML format). OData represents a standard way to query and update data that is

---

3. http://odata.org

TABLE 10.1 OData HTTP Verb Mappings

HTTP Verb	Data Operation
GET	Read
POST	Update
PUT	Insert
DELETE	Delete

Web-friendly. It uses REST-based URI syntax for querying, shaping, filtering, ordering, paging, and updating data across the Internet.

How OData Works

OData is meant to make it simple to handle typical data operations over HTTP. While it is common in Web services to expose different methods for data operations such as Create, Read, Update, and Delete (CRUD), OData takes a different approach. Instead of creating different operations, it leans on the HTTP stack to allow for different HTTP verbs to mean different operations. OData maps HTTP verbs to these CRUD operations, as shown in Table 10.1.

OData supports two different data formats: JSON and AtomPub. These formats are used to communicate in both directions. So if you wanted to insert a record into an OData feed, you would use an HTTP PUT to push a JSON or AtomPub version of a new entity. Each type of entity that you can manipulate using OData is called an **endpoint.** An endpoint is a type of entity that can be queried, inserted, updated, and/or deleted (though not all operations may be permitted). When you navigate to an OData feed it returns a document that tells you about the endpoints it exposes. For example, if you navigate to http://www.nerddinner.com/Services/OData.svc it returns an AtomPub document, like so:

```
<?xml version="1.0" encoding="iso-8859-1" standalone="yes"?>
<service xml:base="http://www.nerddinner.com/Services/OData.svc/"
         xmlns:atom="http://www.w3.org/2005/Atom"
         xmlns:app="http://www.w3.org/2007/app"
         xmlns="http://www.w3.org/2007/app">
```

```
<workspace>
  <atom:title>Default</atom:title>
  <collection href="Dinners">
    <atom:title>Dinners</atom:title>
  </collection>
  <collection href="RSVPs">
    <atom:title>RSVPs</atom:title>
  </collection>
</workspace>
</service>
```

The collection elements in the AtomPub document tell us that the service supports two endpoints (`Dinners` and `RSVPs`). These endpoints have an `href` attribute that points to their name. This link to the endpoint indicates the path to the data. For example, to show the dinners in the feed, you would resolve the `Dinners href` attribute to create a URI such as http://www.nerddinners.com/Services/OData.svc/Dinners.

By navigating to that URI, it returns an AtomPub document that contains all the dinners.

The URI

You may be wondering why it returns an AtomPub document. After all, if OData supports both AtomPub and JSON as data formats, why is the browser returning AtomPub data? OData determines the right type of data to return based on HTTP Accept headers. When an HTTP call is made, a header usually exists to say what kinds of data the recipient can receive. Browsers make their requests with HTML and XML as accepted types and OData detects this and returns XML (AtomPub). If you were to call this in a context such as from JavaScript on an HTML page, the Accept headers would have JSON as an Accept header and OData would then return JSON instead.

The URI syntax says that the service URI can be post-pended with the path to a named endpoint. So both URIs are the path to the `NerdDinner` OData endpoints:

- http://www.nerddinner.com/Services/OData.svc/Dinners
- http://www.nerddinner.com/Services/OData.svc/RSVPs

When you look at the AtomPub data returned by the endpoints, each result is in an element called `entry`:

```
<entry>
  <id>http://www.nerddinner.com/Services/OData.svc/Dinners(1)</id>
  <title type="text" />
  <updated>2011-02-13T00:02:47Z</updated>
  <author>
    <name />
  </author>
  <link rel="edit"
        title="Dinner"
        href="Dinners(1)" />
  <link rel="..."
        type="application/atom+xml;type=feed"
        title="RSVPs"
        href="Dinners(1)/RSVPs" />
  <category term="NerdDinnerModel.Dinner"
            scheme="..." />
  <content type="application/xml">
    <m:properties>
      <d:DinnerID m:type="Edm.Int32">1</d:DinnerID>
      <d:Title>ALT.NERD Dinner</d:Title>
      <d:EventDate m:type="Edm.DateTime">
        2009-02-27T20:00:00
      </d:EventDate>
      <d:Description>
Are you in town for the ALT.NET Conference? Are you a .NET person?
Join us at this free, fun, nerd dinner. Well, you pay for your food!
But, still! Come by Red Robin in Redmond Town Center at 8pm Friday.
      </d:Description>
      <d:HostedBy>shanselman</d:HostedBy>
      <d:ContactPhone>503-766-2048</d:ContactPhone>
      <d:Address>7597 170th Ave NE, Redmond, WA</d:Address>
      <d:Country>USA</d:Country>
      <d:Latitude m:type="Edm.Double">47.670172</d:Latitude>
      <d:Longitude m:type="Edm.Double">-122.1143</d:Longitude>
      <d:HostedById>shanselman</d:HostedById>
    </m:properties>
  </content>
</entry>
```

Although this format is usually hidden from you on the phone, there are a couple of pieces of information in an entry that are of interest. In the entry is a list of links. The first is an "edit" link, which shows the address of the entry on its own. The format takes the form of parentheses with the primary key of the entry. So to retrieve just this entry you could use this link via the relative URI:

```
http://www.nerddinner.com/Services/OData.svc/Dinners(1)
```

The other link that is listed in this example is a related entity link. To retrieve the list of RSVPs for this particular dinner, you could also use the relative URI:

```
http://www.nerddinner.com/Services/OData.svc/Dinners(1)/RSVPs
```

This is at the heart of the relational nature of the OData feed. You can navigate using simple REST-style URIs to get at related entities.

As mentioned earlier, it depends on the HTTP verb during the request as to what the endpoint does with the request. In the browser, all the requests are GET requests, which means they read the data. OData supports a number of query options that allow you to decide how you want to retrieve the data. These query options allow you to specify a query against the endpoint including filtering, sorting, shaping, and paging. For example, to sort the dinners by date you can use the $orderby query option:

```
http://www.nerddinner.com/Services/OData.svc/
Dinners?$orderby=EventDate
```

Table 10.2 shows the supported query options.

TABLE 10.2 OData Query Options

Query Option	Meaning
$orderby	Sorts the results (ascending or descending)
$skip	Seeks into the results before returning results
$top	Limits the results to a set number of results
$filter	Limits the results based on a predicate
$expand	Embeds related entities instead of providing links
$select	Limits the fields returned on an entity

Each of these can be combined to change the result from the OData feed. Let's look at each in the sections that follow.

$orderby

This query option allows you to specify one or more field names (separated by a comma) to use when sorting the results. Each field name can have the suffix "desc" added to mark that the sorting should be done in descending order. Some examples include the following:

```
http://.../Dinners?$orderby=Title
http://.../Dinners?$orderby=Title desc
http://.../Dinners?$orderby=Title,EventDate
http://.../Dinners?$orderby=Title desc,EventDate
http://.../Dinners?$orderby=Title desc,EventDate desc
```

$skip *and* $top

These query options are used to limit and span the number of results in the entire result. The $top query option specifies the maximum number of results to return and the $skip query option is used to specify how many of the results to not return before starting to return results. Although using the $top query option to return only a set number of results is typical on its own, $skip is used almost exclusively with the $top option to provide a paging mechanism. When using $top and $skip for paging, $skip should be preceded by the $top query option so that the records are skipped first and then limited by the number. Otherwise, you will not get the paging you expect. Here are some samples:

```
http://.../Dinners?$top=10
http://.../Dinners?$skip=10&$top=10
```

$filter

The purpose of the $filter query option is to provide a predicate with which to only return results that match the predicate. The language definition for predicates is a robust set of operators and functions. Typically you will need to specify the name of a field to use in the predicate along with an operator and/or functions. The $filter query can specify the name of a simple field such as:

```
http://.../Dinners?$filter=Country eq 'China'
```

In this example, `Country` is the field name followed by an operator (eq means equals) and a value to compare it to. Strings should be delimited by single quotes. Instead of the simple field name you can use navigation to a related entity as well (as long as it is a 1-to-1 relationship). For example:

```
http://.../Suppliers?$filter=Address/City eq 'Atlanta'
```

In this example, the `Supplier` has a property called `Address` that contains a `City` field. So you can filter the `Suppliers` by the city name in this way. Table 10.3 lists the operators and Table 10.4 lists the functions that you can use in a `$filter` query option.

TABLE 10.3 `$filter` Operators

Operator	Description	Example
Logical Operators		
Eq	Equal	/Suppliers?$filter=Address/City eq 'Redmond'
Ne	Not equal	/Suppliers?$filter=Address/City ne 'London'
Gt	Greater than	/Products?$filter=Price gt 20
Ge	Greater than or equal	/Products?$filter=Price ge 10
Lt	Less than	/Products?$filter=Price lt 20
Le	Less than or equal	/Products?$filter=Price le 100
And	Logical and	.../Products?$filter=Price le 200 and Price gt 3.5
Or	Logical or	.../Products?$filter=Price le 3.5 or Price gt 200
Not	Logical negation	.../Products?$filter=not endswith(Description,'milk')

continues

TABLE 10.3 $filter **Operators** (*continued*)

Operator	Description	Example
Arithmetic Operators		
Add	Addition	.../Products?$filter=Price add 5 gt 10
Sub	Subtraction	.../Products?$filter=Price sub 5 gt 10
Mul	Multiplication	.../Products?$filter=Price mul 2 gt 2000
Div	Division	.../Products?$filter=Price div 2 gt 4
Mod	Modulo	.../Products?$filter=Price mod 2 eq 0
Grouping Operator		
( )	Precedence grouping	.../Products?$filter=(Price sub 5) gt 10

TABLE 10.4 $filter **Functions**[1]

Function	Example
String Functions	
bool substringof(string po, string p1)	.../Customers?$filter=substringof('Alfreds' ,CompanyName) eq true
bool endswith(string p0, string p1)	.../Customers?$filter=endswith(CompanyName, 'Futterkiste') eq true
bool startswith(string p0, string p1)	.../Customers?$filter=startswith(CompanyName,'Alfr') eq true
int length(string p0)	.../Customers?$filter=length(CompanyName) eq 19
int indexof(string p0, string p1)	.../Customers?$filter=indexof(CompanyName,' lfreds') eq 1
string replace(string p0, string find, stringreplace)	.../Customers?$filter=replace(CompanyName,' ', '') eq 'AlfredsFutterkiste'

---

1. Function table from the OData Spec: www.odata.org/developers/protocols/uri-conv entions#FilterSystemQueryOption

TABLE 10.4 $filter Functions (*continued*)

Function	Example
string substring(string p0, int pos)	.../Customers?$filter=substring(CompanyName,1) eq 'lfreds Futterkiste'
string substring(string p0, int pos, intlength)	.../Customers?$filter=substring(CompanyName,1, 2) eq 'lf'
string tolower(string p0)	.../Customers?$filter=tolower(CompanyName) eq 'alfreds futterkiste'
string toupper(string p0)	.../Customers?$filter=toupper(CompanyName) eq 'ALFREDS FUTTERKISTE'
string trim(string p0)	.../Customers?$filter=trim(CompanyName)eq 'Alfreds Futterkiste'
string concat(string p0, string p1)	.../Customers?$filter=concat(concat(City, ', '), Country) eq 'Berlin, Germany'
Date Functions	
int day(DateTime p0)	.../Employees?$filter=day(BirthDate)eq 8
int hour(DateTime p0)	.../Employees?$filter=hour(BirthDate)eq 0
int minute(DateTime p0)	.../Employees?$filter=minute(BirthDate)eq 0
int month(DateTime p0)	.../Employees?$filter=month(BirthDate)eq 12
int second(DateTime p0)	.../Employees?$filter=second(BirthDate)eq 0
int year(DateTime p0)	.../Employees?$filter=year(BirthDate)eq 1948
Math Functions	
double round(double p0)	.../Orders?$filter=round(Freight)eq 32
decimal round(decimal p0)	.../Orders?$filter=round(Freight)eq 32
double floor(double p0)	.../Orders?$filter=filter=round(Freight) eq 32
decimal floor(decimal p0)	.../Orders?$filter=floor(Freight)eq 32

continues

TABLE 10.4 `$filter` Functions (*continued*)

Function	Example
`double ceiling(double p0)`	`.../Orders?$filter=ceiling(Freight)eq 33`
`decimal ceiling(decimal p0)`	`.../Orders?$filter=floor(Freight)eq 33`
Type Functions	
`bool IsOf(type p0)`	`.../Orders?$filter=isof(` `'NorthwindModel.Order')`
`bool IsOf(expression p0, type p1)`	`.../Orders?$filter=isof(ShipCountry,` `'Edm.String')`

`$expand`

The purpose of this query option is to allow you to embed specific related data in the results of a request. As you saw earlier, you can access a related entity by following the path to the related entries, like so:

```
http://.../Dinners(1)/RSVPs
```

The `$expand` query option lets you return not only the main endpoint you are making a quest from, but also the related entries in a single call. For example, you can include the RSVPs in the request for dinners like so:

```
http://.../Dinners?$expand=RSVPs
```

This will return the dinners plus any RSVPs for those dinners. If you have a complex chain of related entities such as Customer→Order→ OrderDetails→Products, a single `$expand` query option can include the entire chain by including the path to the deepest part of the object tree:

```
http://.../Customers?$expand=Orders/OrderDetails/Products
```

Finally, you can include multiple expansions by separating individual expansion query options with a comma:

```
http://.../Customers?$expand=Orders/OrderDetails,SalesPeople
```

This request would return the customers and include not only the orders and the details for each order, but also any salespeople for each customer.

$select

This query option is used to limit the fields that the endpoint returns in the request. The $select query option lets you specify a comma-delimited list of fields to return:

```
http://.../Dinners?$select=DinnerID,Title,EventDate
```

Use of this query option tells the OData feed to only return the specified fields:

```
<entry>
  <id>http://www.nerddinner.com/Services/OData.svc/Dinners(1)</id>
  <title type="text"></title>
  <updated>2011-02-13T02:36:44Z</updated>
  <author>
    <name />
  </author>
  <link rel="edit" title="Dinner" href="Dinners(1)" />
  <category term="NerdDinnerModel.Dinner" scheme="..." />
  <content type="application/xml">
    <m:properties>
      <d:DinnerID m:type="Edm.Int32">1</d:DinnerID>
      <d:Title>ALT.NERD Dinner</d:Title>
      <d:EventDate m:type="Edm.DateTime">
        2009-02-27T20:00:00
      </d:EventDate>
    </m:properties>
  </content>
</entry>
```

Note that when you use the $select query option, you are specifying the fields to return for every entity. This impacts when you use it in conjunction with the $expand query option in that you must include the fields you want from the related entities as well. For example, you could use $select to limit the fields from the main endpoint as well as the related entries:

```
/Dinners?$expand=RSVPs&$select=Title,EventDate,RSVPs/AttendeeName
```

This query would result in filtering the main endpoint (Dinners), but would only include the AttendeeName in the related entry (RSVPs).

Using OData on the Phone

One of the unique challenges of the phone is to create really efficient access to data that is supported by services. Public services such as Twitter, eBay, and others have already defined the types of services they are supporting. OData is one of those formats that you may find yourself supporting if you want to consume data from a provider that supports it. You also may decide to expose your own data via OData when you want to expose it to the phone. The main reason OData is a compelling option on the phone is that it allows you to be very specific with how you want to query the data. Being able to tune an application and only return the data (e.g., fields) required using the $select query option means you can be very efficient when accessing data via the phone.

Another reason that using OData with the phone is compelling is that it is supported by a rich client-side model to the data you are dealing with. This is especially helpful if you need to be able to support data modification. The OData client for the phone supports a client-side context object that tracks changed objects for you automatically and allows you to batch those changes back to the server.

OData supports a variety of different platforms (PHP, iPhone, .NET, Java) and support for OData is included natively in the Windows Phone SDK 7.1.

Generating a Service Reference for OData

In the Web Services section, you used Visual Studio to add a service reference, which built an object model for the service. Creating a service reference to an OData feed is accomplished in the same way. You simply use the Add Service Reference option on your project (see Figure 10.1 earlier in this chapter). When the Add Service Reference dialog appears, you insert an OData feed (or click Discover for data services in your own project) as shown in Figure 10.4.

Much like adding a service reference to a Web service, adding a service reference to an OData feed requires a feed location (e.g., http://www. nerddinner.com/Services/OData.svc). When you press the Go button, it retrieves the metadata about the feed to allow you to create the reference classes to the OData feed. Adding this service reference not only creates

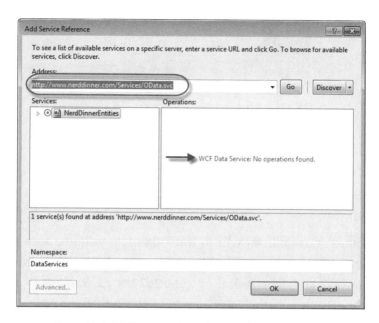

FIGURE 10.4 Adding a service reference to an OData feed

classes for the different entities on the OData feed, but also creates a context class that is used when interacting with the OData feed. By using this context class, you can retrieve and update data in the OData feed (assuming it supports updating). Let's see how that is accomplished.

Retrieving Data

To start accessing data with OData, you will need to create an instance of the context class. This class gives you access to the different endpoints in the service. Depending on the service, this class name could end in "Entities" or "Context". For example, in the NerdDinner OData service, it is called `NerdDinnerEntities`. This context class (and the other classes generated by the service reference) are contained in a namespace that was specified by the OData service. Figure 10.5 shows adding the `Data-Services` namespace to have access to the entity class. The name of this namespace was specified in the Add Service Reference dialog (shown in Figure 10.4).

Creating an instance of the context class requires that you include the address of the service you are querying, like so:

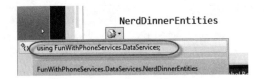

FIGURE **10.5 Adding a** using **statement to the data service**

```
// The Service Address
var svcUri = new Uri("http://www.nerddinner.com/Services/OData.svc");

// Create the Context
NerdDinnerEntities ctx = new NerdDinnerEntities(svcUri);
```

This context class allows you to track the date you get back from the server, but the DataServiceCollection class is at the center of how you will query data. The DataServiceCollection class is a generic collection that supports the INotifyCollectionChanged interface so that it easily supports data binding. You effectively can ask the collection to load itself with a query as shown here:

```
// Craft the Query
var query = from d in ctx.Dinners
            where d.HostedBy == "Shawn Wildermuth"
            orderby d.Title
            select d;

// Create a collection to hold the results
DataServiceCollection<Dinner> results = new
  DataServiceCollection<Dinner>();

// Handle the completed event to do error handling
results.LoadCompleted += (s, e) =>
  {
    if (e.Error != null)
    {
      MessageBox.Show("An error occurred");
    }
  };

// Load the Collection using the Query
results.LoadAsync(query);

// Bind it to the UI
theList.ItemsSource = results;
```

Although you could craft the query directly via the URI syntax, the OData libraries for the phone allow you to use LINQ to describe your query, which it automatically converts to the URI query syntax for you. Next, the code creates a new instance of the `DataServiceCollection` that expects it to be filled with `Dinner` objects. Then it handles the collection's `LoadCompleted` event to handle any errors. And finally it calls `LoadAsync` to actually start the operation asynchronously. At that point you can bind the collection to the user interface, and the query will execute and fill in the `DataServiceCollection` in the background. Once the collection is filled, the data binding will take over and display the results.

If you are using projections in your queries, the results will continue to be the full objects but only the requested fields will be filled in. For example, if you request only some of an entity's properties, it will return just those properties as shown here:

```
// Craft the Query
var query = from d in ctx.Dinners
            where d.Country == "USA"
            orderby d.Title
            select new Dinner()
            {
              DinnerID = d.DinnerID,
              Title = d.Title,
              HostedBy = d.HostedBy
            };
```

If you are using expansion in your queries, the results will continue to be the main objects you've requested, but the navigated objects will be serialized as part of your query. For example, adding an expansion (via the `Expand` method) will eager-load any related entities:

```
// Craft the Query
var query = from d in ctx.Dinners.Expand("RSVPs")
            where d.Country == "USA"
            orderby d.Title
            select d;
```

Updating Data

Querying the OData feed via the `DataServiceCollection` and context classes allows you to track any changes to the collection so that as objects are changed that exist in the `DataServiceCollection`, those objects are marked as changed. In addition, if you want to delete an item you can simply remove it from the `DataServiceCollection`. Removing the item from the collection marks it as a deleted element. And finally, you can add items to the collection to mark them as new (or inserted) objects. You can see examples of this below:

```
// Changing items from the collection marks them as changed
Dinner dinner = _dinnerCollection[0];
dinner.EventDate = DateTime.Today;

// Adding Items marks them as New (Inserted)
Dinner newDinner = new Dinner()
{
  EventDate = DateTime.Today,
    HostedBy = "ShawnWildermuth"
};
_dinnerCollection.Add(newDinner);

// Removing Items marks them as Deleted
Dinner oldDinner = _dinnerCollection[1];
_dinnerCollection.Remove(oldDinner);
```

As changes are tracked, you can update those changes with the server (assuming you have permission to update the server) by calling `Begin-SaveChanges` on the context object. Remember, even though the `Data-ServiceCollection` is the holder of the objects, ultimately it's the context object that actually tracks them so that when you call `Begin-SaveChanges` it will batch these changes back to the server, like so:

```
// Save the changes
_theContext.BeginSaveChanges(new AsyncCallback(r =>
  {
    // Response is a collection of errors
    DataServiceResponse response = _theContext.EndSaveChanges(r);

    // If there are any errors, show message
    if (response.Any(op => op.Error != null))
    {
      MessageBox.Show("Sorry, update failed.");
    }
  }), null);
```

Unlike reading data using OData, the saving of changes uses an older asynchronous style using an `AsyncCallback` object. The code can be simplified (as shown here) by using a lambda to handle the changes inline, but it still requires the `BeginSaveChanges` and `EndSaveChanges` methods to be called. The `DataServiceResponse` object that is returned at the end of the save operation is a collection of `OperationResponse` objects. These `OperationResponse` objects contain the exception that was encountered on the server. In this example, the LINQ `Any` operator checks to see if any of the responses have an error (which is the `Exception`). If the collection is empty or responses do not contain errors, you can assume that the operation completed successfully.

> **▪ OData and Transactions**
>
> Updates to an OData service are handled in a transactional way so that if any errors were found you can assume the complete update failed.

Using Push Notifications

At times you will want to keep the user apprised of some change that happen on the server. While you could write a background agent (as shown in Chapter 9, Multitasking), making network requests every 30 minutes may not be often enough. You may also decide that you want to determine when to update a phone remotely (as close to the data as possible). So instead, you will want to use the phone's ability to receive messages from the cloud. These messages are called **push notifications.**

Push notifications allow your application to register a particular phone to receive these notifications. Once the phone is registered, you can send messages from an Internet-connected server to Microsoft's Push Notification Service. This service is hosted by Microsoft and allows your messages to reach users' phones (and deals with the hard problems of out-of-range and powered-off phones). Using push notifications does require that you have a way to communicate with a service that you are responsible for; typically this is a Web service hosted on a public Internet server. You can see the general flow of push notifications in Figure 10.6.

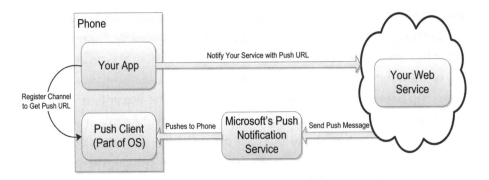

FIGURE 10.6 **Push notification message flow**

Using push notifications requires several steps, but they are all fairly straightforward.

1. Your app opens a push notification channel.
2. Your app receives a special URL that can be used to send messages to this specific phone.
3. Your app sends that URL to a service that wishes to send you notifications.
4. Sometime later, the service can post a new message to Microsoft's Push Notification Service using the special URL from the phone.
5. The Push Notification Service finds your phone and sends the message to the phone.

Push notifications can be used to do three tasks on the phone: alert the user with a toast notification, update a Live Tile, and send your application raw messages. You will see how to do all three, but let's start with the basics so that you can learn how all the moving parts work together.

Push Notification Requirements

Using push notifications comes with several requirements. To use push notifications you must do the following.

- Add the push notification requirement to your WMAppManifest.xml file:

```
...
<Capabilities>
  <Capability Name="ID_CAP_PUSH_NOTIFICATION"/>
</Capabilities>
...
```

- Alert the user in the user interface of the application (usually during the first launch of your application) that you are using push notifications.
- Allow the user to disable push notifications.

You can only have one push notification channel per application. In addition, only 15 applications on a phone can use push notifications. When you first open your HTTP notification channel you may receive an error message stating that the channel could not be opened. This is typically thrown when there are too many push notification applications on the phone.

Preparing the Application for Push Notifications

To get started, you will need an instance of a class that will register your application for push notifications as well as listen for certain events (e.g., notifications, errors, etc.). This class is called `HttpNotification-Channel`. When you launch your application the first time, you will register this instance of the application (specifically the application on this particular phone) with the Push Notification Service by creating a new instance of the class and passing in a channel name that is specific to the application (e.g., the app name):

```
HttpNotificationChannel  myChannel =
  new HttpNotificationChannel("MYAPP");
```

The name of the channel is usually the name of the application, but since you can only have one channel, it is not important that this name be unique to the phone or user. On subsequent launches of your application, you can get the channel by calling the static `Find` method using the name like so:

```
HttpNotificationChannel  myChannel =
  HttpNotificationChannel.Find("MYAPP");
```

In practice, determining if this is the first invocation or not can be tedi-
ous, so you should just try to find the channel, and if it is not found just
create it like so:

```
public class PushSignup
{
  const string CHANNELNAME = "FunWithToast";
  private HttpNotificationChannel _myChannel = null;

  public void CreateChannel()
  {
    try
    {
      if (_myChannel == null)
      {
        // Attempt to get the channel
        _myChannel = HttpNotificationChannel.Find(CHANNELNAME);

        // Can't Find the Channel, so create it
        if (_myChannel == null)
        {
          _myChannel = new HttpNotificationChannel(CHANNELNAME);
  ...
```

This way, you won't need to worry about keeping state on the first invocation of your application.

Once you have a valid channel, that channel will give your server access
to a special URI that is used to send push notifications. Again, the channel
is a little different, depending on whether it was newly created or located
with the Find method. You would think that you should be able to take
the newly created channel and retrieve the URI, but that only works on
existing channels (i.e., channels located with Find):

```
  ...

  // If the channel uri is valid now, then send it to the server
  // Otherwise it'll be registered when the URI is updated
  if (_myChannel.ChannelUri != null)
  {
    // Use the ChannelUri to tell your server about the new phone
  }

  ...
```

Newly created channels will get the URI asynchronously. To know
when the URI is available, you need to register the ChannelUriUpdated

event on the channel object. When this event is fired you can use the `Chan-nelUri`. In practice you should handle this in both cases to ensure that any changes to the URI are caught and reported:

```
_myChannel.ChannelUriUpdated += myChannel_ChannelUriUpdated;

...

void myChannel_ChannelUriUpdated(object sender,
                                 NotificationChannelUriEventArgs e)
{
  // Use the ChannelUri to tell your server about the new phone
}
```

Once you have the channel open and a way to get the `ChannelUri`, you're ready to set up the server side of the equation.

Setting Up the Server for Push Notifications

Because push notifications come from the server (and are not necessarily triggered by a running application on the phone), you need some code somewhere on a server that is reachable by the phone. The reasoning for this is that the server uses the channel URI to send any push notifications to the phone. How you choose to send that information to the server is up to you, but a common approach is to use a simple Web service.

> **▪ Server-Side Code**
>
> There is nothing special about the server-side code that requires the server technology and operating system to be Microsoft servers or ASP.NET, but this example will use WCF and ASP.NET since that is a common approach. You can use whatever server technology and operating systems you choose. This chapter does not go into detail about how WCF or ASP.NET works. If you are not proficient in .NET server-side technologies, you will need an additional reference for those technologies.

For example, we can have an ASP.NET Web project that hosts a WCF service that accepts the channel URI from the phone. In this example, I've set up a simple database to hold the channel URIs as well as a WCF service that can accept the channel URIs. The WCF service interface looks simple enough:

```
[ServiceContract]
public interface IMyPhoneService
{
  [OperationContract]
  void RegisterPhoneApp(string channelUrl);
}
```

The expectation is that the phone will call a service method with the channel URI when it receives it so that the server has a list of people to notify. It would be common to include additional information, but for this example let's keep it simple. The implementation of the service would look like this:

```
public void RegisterPhoneApp(string channelUrl)
{
  // Add to list of URIs to notify
  using (var ctx = new PhoneEntities())
  {
    if (ctx.Phones.Where(p => p.PhoneUrl == channelUrl).Count() == 0)
    {
      var phone = Phone.CreatePhone(0, channelUrl, DateTime.Now);
      ctx.Phones.AddObject(phone);
      ctx.SaveChanges();
    }
  }
}
```

In this case the code just uses the Entity Framework to store the channel URI in the database for use later. This means the phone code has to change to actually use this Web service. To do this you can just create a small helper method that calls the service:

```
void SendChannelUriToService(string uri)
{
  var client = new MyPhoneServiceClient();

  // Handle the completed event
  client.RegisterPhoneAppCompleted += (s, e) =>
    {
      if (e.Error != null)
      {
        // Log Error (User Can't Fix this error)
      }
    };

  // Send the channel URI to the server
  client.RegisterPhoneAppAsync(uri);
}
```

This code should look familiar because it is just a Web service call similar to what we discussed earlier in the chapter. Now that you have the helper method, you can call this when you have the channel URI:

```
public void CreateChannel()
{
  try
  {
    if (_myChannel == null)
    {
      // Attempt to get the channel
      _myChannel = HttpNotificationChannel.Find(CHANNELNAME);

      // Can't Find the Channel, so create it
      if (_myChannel == null)
      {
        _myChannel = new HttpNotificationChannel(CHANNELNAME);

        // Add Event Handlers
        _myChannel.ChannelUriUpdated += myChannel_ChannelUriUpdated;

        // Open the channel since it's a new channel
        _myChannel.Open();
      }
      else
      {
        // Add Event Handlers
        _myChannel.ChannelUriUpdated += myChannel_ChannelUriUpdated;
      }

      // If the channel uri is valid now, then send it to the server
      // Otherwise it'll be registered when the URI is updated
      if (_myChannel.ChannelUri != null)
      {
        // Use the ChannelUri to tell your server about the new phone
        SendChannelUriToService(_myChannel.ChannelUri.ToString());
      }
    }

    return;
  }
  catch (Exception)
  {
    MessageBox.Show("Failed to create Push Notification Channel");
  }
}

...
```

```
void myChannel_ChannelUriUpdated(object sender,
                              NotificationChannelUriEventArgs e)
{
  // Use the ChannelUri to tell your server about the new phone
  SendChannelUriToService(e.ChannelUri.ToString());
}
```

With the channel URI in hand, you are ready to push notification messages to the phone!

■ Authenticating Push Notifications

The Push Notification Service can use a custom certificate that you send to Microsoft during the submission process for authenticating your phone. Push notifications are specifically from your own server(s). The documentation covers this in detail and is beyond the scope of this book.

Raw Notifications

The first type of push notification you can send is a raw notification. This type of message is sent directly to a running application. If your application isn't running when you send a raw notification, it is just dropped. Raw notifications allow you to send any sort of message to your application.

To send a raw notification you have to use an HTTP call to POST a message to Microsoft's Push Notification Service. The messages you send to this service are simple HTTP calls and are not using Web services. So you can write your server code to send a message using the HttpWebRequest and HttpWebResponse classes. First you would set up your request like so:

```
HttpWebRequest request =
  (HttpWebRequest)WebRequest.Create(pushUri);

// For Raw Update use type of body
// (text/plain or text/xml are typical)
request.ContentType = "text/plain";

// Specify the HTTP verb (must be POST)
request.Method = "POST";
```

This creates the Web request, sets the content type of the message (usually a string or XML fragment), and specifies that the request is posting data. Next you need to add some headers to make the push notification message make sense to the Push Notification Service:

```
// Use a generated unique ID to prevent duplicate push messages
request.Headers.Add("X-MessageID", Guid.NewGuid().ToString());

// Send Raw Immediate requires class == 3
request.Headers.Add("X-NotificationClass", "3");
```

To ensure that the message doesn't get duplicated it's a good idea to add a message ID header with a unique ID. The X-Notification-Class header is required for the Push Notification Service to know how to handle the request. This header needs to be "3" for a raw request. The X-NotificationClass header specifies the type of message ("3" is a raw notification). This tells the Push Notification Service to send the message immediately. It also allows you to add 10 to the value if you want to post it to be sent in 450 seconds (7.5 minutes), or you can add 20 to send it in 900 seconds (15 minutes). So, for a raw notification, the valid values are 3, 13, and 23, respectively. Next you will need to push the message into the request stream to have it become part of the message:

```
// Send it
byte[] notificationMessage = Encoding.UTF8.GetBytes(message);
request.ContentLength = notificationMessage.Length;
using (Stream requestStream = request.GetRequestStream())
{
    requestStream.Write(notificationMessage,
                        0,
                        notificationMessage.Length);
}
```

First the message is encoded into UTF-8 and converted to a byte array so that you can stuff it into the request stream. Once the size of the message is known, you can specify the length of the message (ContentLength). Then you can use the stream to push the message into the request. Finally, you are ready to send the message:

```
// Sends the notification and gets the response.
try
{
    HttpWebResponse response = (HttpWebResponse)request.GetResponse();

    return HandleResponse(response);
}
```

```
catch (Exception ex)
{
  return string.Concat("Exception during sending message",
ex.Message);
}
```

By calling the `GetResponse` method, you will retrieve an `HttpWeb-Response`, which will tell you about the status of the message you just posted to the server. Here is the entire example for pushing a raw notification to a phone:

```
string SendRawMessage(string pushUri, string message)
{
  HttpWebRequest request =
    (HttpWebRequest)WebRequest.Create(pushUri);

  // For Raw Update use type of body
  // (text/plain or text/xml are typical)
  request.ContentType = "text/plain";

  // Specify the HTTP verb (must be POST)
  request.Method = "POST";

  // Use a generated unique ID to prevent duplicate push messages
  request.Headers.Add("X-MessageID", Guid.NewGuid().ToString());

  // Send Raw Immediate requires class == 3
  request.Headers.Add("X-NotificationClass", "3");

  // Send it
  byte[] notificationMessage = Encoding.UTF8.GetBytes(message);
  request.ContentLength = notificationMessage.Length;
  using (Stream requestStream = request.GetRequestStream())
  {
    requestStream.Write(notificationMessage,
                        0,
                        notificationMessage.Length);
  }

  // Sends the notification and gets the response.
  try
  {
    HttpWebResponse response = (HttpWebResponse)request.GetResponse();

    return HandleResponse(response);
  }
  catch (Exception ex)
  {
```

```
      return string.Concat("Exception during sending message",
                            ex.Message);
  }
}
```

The response from the server takes the form of two different pieces of information that are handled in the `HandleResponse` method:

```
string HandleResponse(HttpWebResponse response)
{
  // Pull status from headers if they exist
  string notificationStatus =
    response.Headers["X-NotificationStatus"];
  string deviceStatus = response.Headers["X-DeviceConnectionStatus"];
  string subscriptionStatus =
    response.Headers["X-SubscriptionStatus"];

  switch (response.StatusCode)
  {
    case HttpStatusCode.OK: // 200
      {
        //
        return "Success";
      }
    case HttpStatusCode.BadRequest: // 400
      {
        return "Failed, bad request";
      }
    case HttpStatusCode.Unauthorized: // 401
      {
        return "Not authorized";
      }
    case HttpStatusCode.NotFound: // 404
      {
        return "Not Found";        }
    case HttpStatusCode.MethodNotAllowed: // 405
      {
        return "Only POST Allowed";
      }
    case HttpStatusCode.NotAcceptable: // 406
      {
        return "Request Not Acceptable";
      }
    case HttpStatusCode.PreconditionFailed: // 412
      {
        return "Failed to Meet Preconditions";
      }
    case HttpStatusCode.ServiceUnavailable: // 503
```

```
        {
          return "Service down";
        }


    }

    return "Success";
  }
```

First there are three headers that the service returns with the response to indicate certain status information to help you determine what the service actually did. Table 10.5 details these headers.

These header response values will mean different things depending on the HTTP status code that was returned. Table 10.6 shows the mix of HTTP status codes and these header values (from the Windows Phone OS 7.1 documentation).

Using the information you can glean from the HTTP status and the headers, you can determine what to do with the message. This example does not show any retry mechanism and simply loses the notification if the message fails. Depending on the value of the message you may choose what retry level you need to accomplish on the server.

■ SOCKS Proxies

If your phone attaches to a network (e.g., Wi-Fi) that requires a SOCKS proxy to get out to the Internet, push messages will not work.

TABLE 10.5 Push Notification Response Headers

Value	Description
X-NotificationStatus	The status of the notification that was attempted. Valid values include `Received`, `Dropped`, and `QueueFull`.
X-DeviceConnectionStatus	The status of the device the notification attempted to push your message to. Valid values include `Connected`, `InActive`, `Disconnected`, and `TempDisconnected`.
X-SubscriptionStatus	The status of the channel that the phone created. Valid values include `Active` and `Expired`.

TABLE 10.6 Response Codes and Header Status Codes

Response Code	NotificationStatus	DeviceConnectionStatus	SubscriptionStatus	Comments
200 OK	Received	Connected	Active	The notification request was accepted and queued for delivery.
200 OK	Received	Temporarily Disconnected	Active	The notification request was accepted and queued for delivery. However, the device is temporarily disconnected.
200 OK	QueueFull	Connected	Active	Queue overflow. The Web service should resend the notification later. A best practice is to use an exponential backoff algorithm in minute increments.
200 OK	QueueFull	Temporarily Disconnected	Active	Queue overflow. The Web service should resend the notification later. A best practice is to use an exponential backoff algorithm in minute increments.
200 OK	Suppressed	Connected	Active	The push notification was received and dropped by the Push Notification Service. The Suppressed status can occur if the notification channel was configured to suppress push notifications for a particular push notification class.

continues

TABLE 10.6 Response Codes and Header Status Codes (*continued*)

Response Code	NotificationStatus	DeviceConnectionStatus	SubscriptionStatus	Comments
200 OK	Suppressed	Temp Disconnected	Active	The push notification was received and dropped by the Push Notification Service. The Suppressed status can occur if the notification channel was configured to suppress push notifications for a particular push notification class.
400 BadRequest	N/A	N/A	N/A	This error occurs when the Web service sends a notification request with a bad XML document or malformed notification URI.
401 Unauthorzed	N/A	N/A	N/A	Sending this notification is unauthorized. This error can occur for one of the following reasons. There is a mismatch between the subject name of the certificate on the Web service and the subject name of the certificate on the Push Notification Service. The token has been modified. The token is not valid for its subscription.
404 Not Found	Dropped	Connected	Expired	The subscription is invalid and is not present on the Push Notification Service. The Web service should stop sending new notifications to this subscription, and drop the subscription state for its corresponding application session.

TABLE 10.6 Response Codes and Header Status Codes (*continued*)

Response Code	NotificationStatus	DeviceConnectionStatus	SubscriptionStatus	Comments
404 Not Found	Dropped	Temporarily Disconnected	Expired	The subscription is invalid and is not present on the Push Notification Service. The Web service should stop sending new notifications to this subscription, and drop the subscription state for its client.
404 Not Found	Dropped	Disconnected	Expired	The subscription is invalid and is not present on the Push Notification Service. The Web service should stop sending new notifications to this subscription, and drop the subscription state for its client.
405 Method Not Allowed	N/A	N/A	N/A	Invalid method (PUT, DELETE, CREATE). Only POST is allowed when sending a notification request.
406 Not Acceptable	Dropped	Connected	Active	This error occurs when an unauthenticated Web service has reached the per-day throttling limit for a subscription. The Web service can try to resend the push notification every hour after receiving this error. The Web service may need to wait up to 24 hours before normal notification flow will resume.

continues

Table 10.6 Response Codes and Header Status Codes (*continued*)

Response Code	NotificationStatus	DeviceConnectionStatus	SubscriptionStatus	Comments
406 Not Acceptable	Dropped	Temp Disconnected	Active	This error occurs when an unauthenticated Web service has reached the per-day throttling limit for a subscription. The Web service can try to resend the push notification every hour after receiving this error. The Web service may need to wait up to 24 hours before normal notification flow will resume.
412 Precondition Failed	Dropped	Inactive	N/A	The device is in an inactive state. The Web service may reattempt sending the request one time per hour at maximum after receiving this error. If the Web service violates the maximum of one reattempt per hour, the Push Notification Service will deregister or permanently block the Web service.
503 Service Unavailable	N/A	N/A	N/A	The Push Notification Service is unable to process the request. The Web service should resend the notification later. A best practice is to use an exponential backoff algorithm in minute increments.

Now that you know how to send the message to the phone, it's time to discuss how the phone reacts to the message. The only change in the phone code to receive raw notifications is to handle the `HttpNotification-Received` event to receive the specific type of message:

```
_myChannel.HttpNotificationReceived +=
  channel_HttpNotificationReceived;
```

This event is fired when the application receives a raw notification:

```
void channel_HttpNotificationReceived(object sender,
                                      HttpNotificationEventArgs e)
{
  StreamReader rdr = new StreamReader(e.Notification.Body);
  string msg = rdr.ReadToEnd();

  Deployment.Current.Dispatcher.BeginInvoke(() =>
  {
    MessageBox.Show(string.Concat("Raw received while app running:",
                                  msg));
  });
}
```

The event sends an object in the `HttpNotificationEventArgs` object that contains information about the notification, including the body of the message as well as the headers. Depending on how you intend to use the raw message, you may add your information as the body of the message or as headers. The event here gives you access to either. It is entirely application-specific how you use raw notifications.

▪ Debugging Push Notifications

Debugging both the phone and the ASP.NET server-side code requires that you set up your solution to start both projects. You can do this via the Solution Properties as shown in Figure 10.7.

Sending Toast Notifications

Toast notifications are simple messages that appear on the top of the phone to alert the user with a specific message from the application. If the user taps the toast message, the application that notified the user will launch.

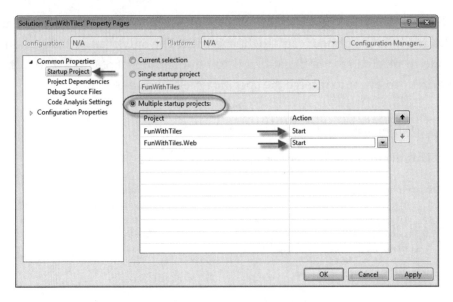

FIGURE 10.7 **Debugging push notifications**

Toast messages are only shown when your application is not currently running. They provide a convenient way to get the user to go to your application and see new or updated information. You can see a sample toast message in Figure 10.8.

For your application to accept toast messages you must tell the phone's shell service that you want to accept toast notifications. You do this by adding this code to the phone's channel registration:

```
// Make sure that you can accept toast notifications
if (!_myChannel.IsShellToastBound) _myChannel.BindToShellToast();
```

Because toast messages are typically received while the application is not running, they are usually used to alert users of certain information or to alert them to start the application. You might also want to be notified

FIGURE 10.8 **A toast message**

when a toast message is sent while your application is running. You can do this by handling the ShellToastNotificationReceived event on the channel:

```
_myChannel.ShellToastNotificationReceived +=
  channel_ShellToastNotificationReceived;
```

This event allows you to be alerted if a toast message comes in while the application is running:

```
void channel_ShellToastNotificationReceived(object sender,
                                            NotificationEventArgs e)
{
  Deployment.Current.Dispatcher.BeginInvoke(() =>
    {
      MessageBox.Show("Toast received while app running");
    });
}
```

◾ Toast Notifications

If a toast message comes in while the application is running, the toast notification will not appear. Toast notifications only appear when your application is not currently in the foreground.

Back on the server, creating these types of messages is very similar to creating a raw notification, but the format of the toast message is very specific. The message you send must be a small XML document that contains information about what to show in the toast message:

```
<?xml version="1.0" encoding="utf-8"?>
<wp:Notification xmlns:wp="WPNotification">
  <wp:Toast>
    <wp:Text1>Title</wp:Text1>
    <wp:Text2>Message</wp:Text2>
  </wp:Toast>
</wp:Notification>
```

The message must look exactly like this, except the first (bolded) part of the message will be in the Text1 section and the second part (unbolded) will be in the Text2 section. The code to send this message is the same as a raw message, but with some minor changes:

```
string SendToastMessage(string pushUri, string message)
{
  HttpWebRequest request =
    (HttpWebRequest)WebRequest.Create(pushUri);

  // For Raw Update use type of body
  // (text/plain or text/xml are typical)
  request.ContentType = "text/xml";

  // Specify the HTTP verb (must be POST)
  request.Method = "POST";
/
  // Use a generated unique ID to prevent duplicate push messages
  request.Headers.Add("X-MessageID", Guid.NewGuid().ToString());

  // Send Toast Immediate requires class == 2
  request.Headers.Add("X-NotificationClass", "2");

  // Specify that the message is a Toast message
  request.Headers.Add("X-WindowsPhone-Target", "toast");

  // Create the XML of the Toast Message
  string toastMessage = @"<?xml version=""1.0"" encoding=""utf-8""?>
                      <wp:Notification
                          xmlns:wp=""WPNotification"">
                        <wp:Toast>
                          <wp:Text1>{0}</wp:Text1>
                          <wp:Text2>{1}</wp:Text2>
                        </wp:Toast>
                      </wp:Notification>";
  string toastXml = string.Format(toastMessage,
                              "This is Toast",
                              message);

  // Send it
  byte[] notificationMessage = Encoding.UTF8.GetBytes(toastXml);
  request.ContentLength = notificationMessage.Length;
  using (Stream requestStream = request.GetRequestStream())
  {
    requestStream.Write(notificationMessage,
                      0,
                      notificationMessage.Length);
  }

  // Sends the notification and gets the response.
  try
  {
    HttpWebResponse response =
      (HttpWebResponse)request.GetResponse();
```

Using Push Notifications ■

```
    return HandleResponse(response);
  }
  catch (Exception ex)
  {
    return string.Concat("Exception during sending message",
                         ex.Message);
  }
}
```

Because the toast message is XML, you must change the content type to `text/xml`. The `X-NotificationClass` value for toast messages is 2 (instead of 3 for raw messages). Like raw messages, you can add 10 and 20 to increase the time before the message is sent (making 2, 12, and 22 the three values for the `X-NotificationClass` header). As it is a toast message, you must add a new header called `X-WindowsPhone-Target` and set it to `toast` to indicate it's a toast message. Finally, the format of the message is different, so you need to create the XML message and you can upload it like any other string.

Creating Live Tiles

In Chapter 6, Developing for the Phone, you learned that the WMManifest.xml file contains the images for the icon used on the list applications for the phone and for applications pinned to the home screen. When a user pins your application to her home screen you can (using push notifications) update the tile. These tiles are called Live Tiles.

Pinned tiles consist of multiple layers, as shown in Figure 10.9.

The layers of the tile correspond to the items in the WMManifest.xml file, as explained in Chapter 6:

```
<?xml version="1.0" encoding="utf-8"?>
<Deployment xmlns="..."
            AppPlatformVersion="7.0">
  <App ...>
    ...
    <Tokens>
      <PrimaryToken TokenID="FunWithToastToken"
                    TaskName="_default">
        <TemplateType5>
          <BackgroundImageURI IsRelative="true"
                              IsResource="false">
          Background.png
          </BackgroundImageURI>
```

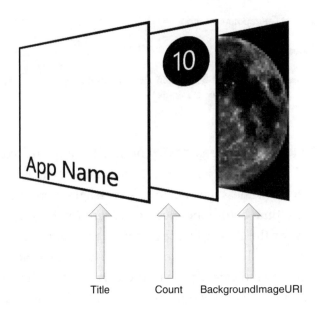

FIGURE 10.9 Tile layers

```
        <Count>0</Count>
        <Title>FunWithToast</Title>
      </TemplateType5>
    </PrimaryToken>
  </Tokens>
</App>
</Deployment>
```

Sending a Live Tile notification is similar to sending a raw notification, but first you must bind your channel to the shell via the BindToShell-Time (after first checking to be sure the channel hasn't already been bound):

```
// Bind to the Shell Tile
if (!_myChannel.IsShellTileBound)
{
  _myChannel.BindToShellTile();
}
```

Binding it to the shell in this way tells the phone to expect Live Tile notifications. Over on the server side, the code is similar to raw push notifications, but a couple of small changes need to occur. The body of the Live Tile update must be in the format of a specific XML document:

```
<?xml version=""1.0"" encoding=""utf-8""?>
<wp:Notification xmlns:wp=""WPNotification"">
  <wp:Tile>
    <wp:BackgroundImage>URL</wp:BackgroundImage>
    <wp:Count>0</wp:Count>
    <wp:Title>Title</wp:Title>
    <wp:BackBackgroundImage>URL</wp:BackBackgroundImage>
    <wp:BackTitle>Title</wp:BackTitle>
    <wp:BackContent>Content</wp:BackContent>
  </wp:Tile>
</wp:Notification>
```

The format of this file is similar to the WMApplication.xml file above. The three layers are all sent to the phone in the form of the push notification and the data in the XML document is used to update the tile. You can also include information for the back of the tile as shown. The additional fields in this format correspond to the back of the Live Tiles, as explained in Chapter 7, Phone Integration.

Once you create this XML document, you are ready to push the change to the phone. So when you push this update from the server you need to make a couple of small changes from the raw push notifications:

```
string SendTileMessage(string pushUri, string imageUrl)
{
  HttpWebRequest request =
    (HttpWebRequest)WebRequest.Create(pushUri);

  // For Raw Update use type of body
  // (text/plain or text/xml are typical)
  request.ContentType = "text/xml";

  // Specify the HTTP verb (must be POST)
  request.Method = "POST";

  // Use a generated unique ID to prevent duplicate push messages
  request.Headers.Add("X-MessageID", Guid.NewGuid().ToString());

  // Send Tile Immediate requires class == 1
  request.Headers.Add("X-NotificationClass", "1");

  // Specify that the message is a Toast message
  request.Headers.Add("X-WindowsPhone-Target", "token");

  // Create the XML of the Tile Message
  string tileMessage = @"<?xml version=""1.0"" encoding=""utf-8""?>
  <wp:Notification xmlns:wp=""WPNotification"">
    <wp:Tile>
```

```
        <wp:BackgroundImage>{0}</wp:BackgroundImage>
        <wp:Count>0</wp:Count>
        <wp:Title>Fun With Push</wp:Title>
        <wp:BackTitle>Back of Tile</wp:BackTitle>
        <wp:BackContent>More Content</wp:BackContent>
    </wp:Tile>
</wp:Notification>";

string tileXml = string.Format(tileMessage, imageUrl);

// Send it
byte[] notificationMessage = Encoding.UTF8.GetBytes(tileXml);
request.ContentLength = notificationMessage.Length;
using (Stream requestStream = request.GetRequestStream())
{
  requestStream.Write(notificationMessage,
                      0,
                      notificationMessage.Length);
}

// Sends the notification and gets the response.
try
{
  HttpWebResponse response =
    (HttpWebResponse)request.GetResponse();

  return HandleResponse(response);
}
catch (Exception ex)
{
  return string.Concat("Exception during sending message",
                       ex.Message);
}
}
}
```

The first change is to make sure the `ContentType` is set to `text/xml` since the body will be the XML document you're going to create. Next the `X-NotificationClass` header needs to be set to `1` for tile updates. Like the raw and toast notifications, you can delay the delivery of the notification by adding 10 or 20, making the valid `X-NotificationClass` values `1`, `11`, and `21` for Live Tile updates. Next you need to add a new header for the `X-WindowsPhone-Target` header and specify its value as `token` to tell the Push Notification Service that this is a tile update. Lastly, you simply need to format the XML message as described earlier. Pushing this to the phone will allow you to update the tile on the phone (if it's pinned to the home screen).

It's a common practice to send an Internet-accessible URI for the `BackgroundImage` value. The image must be exactly 173 × 173 pixels in size and be a 24-bit .png file.

If you need to update a secondary tile, you must include a property of the tile called `Id` that has the URI used for the secondary tile:

```
// Create the XML of the Tile Message
string tileMessage = @"<?xml version=""1.0"" encoding=""utf-8""?>
<wp:Notification xmlns:wp=""WPNotification"">
  <wp:Tile Id="/views/AnotherView.xaml?flightNo=465">
    <wp:BackgroundImage>{0}</wp:BackgroundImage>
    <wp:Count>0</wp:Count>
    <wp:Title>Fun With Push</wp:Title>
    <wp:BackTitle>Back of Tile</wp:BackTitle>
    <wp:BackContent>More Content</wp:BackContent>
  </wp:Tile>
</wp:Notification>";
```

This `Id` attribute must match the URI of the secondary tile exactly; otherwise, it will not update the tile and just ignore the request.

Handling Push Notification Errors

When working with push notifications, you should also handle the `Error-Occurred` event to deal with any exceptional conditions. You should always handle this event as it will help you locate badly formatted messages and low-battery issues that will stop push messages from working. You can handle this event like any other event:

```
_myChannel.ErrorOccurred += channel_ErrorOccurred;
```

The event passes in an error argument that has an `ErrorType` property that can be used to determine the type of error. You should change the way you alert the user to this based on the type of error. In addition, the handler should use the `Dispatcher` to make sure any reporting of these errors occurs on the UI thread (as the `ErrorOccurred` event could be thrown on any thread).

```
void channel_ErrorOccurred(object sender,
                           NotificationChannelErrorEventArgs e)
{
  Deployment.Current.Dispatcher.BeginInvoke(() =>
  {
    switch (e.ErrorType)
```

```
    {
      case ChannelErrorType.ChannelOpenFailed:
        MessageBox.Show("Failed to open channel");
        break;
      case ChannelErrorType.NotificationRateTooHigh:
        MessageBox.Show("Push Notifications are too frequent.");
        break;
      case ChannelErrorType.PayloadFormatError:
        MessageBox.Show(@"XML or Headers were incorrect
                          for the Push Message");
        break;
      case ChannelErrorType.MessageBadContent:
        MessageBox.Show("Live Tile Data is invalid.");
        break;
      case ChannelErrorType.PowerLevelChanged:
        {
          MessageBox.Show(@"Push Notifications are
                            affected by the current power level");

          // Can get the power level from the event info
          ChannelPowerLevel level =
            (ChannelPowerLevel)e.ErrorAdditionalData;

          switch (level)
          {
            case ChannelPowerLevel.NormalPowerLevel:
              // All Push Messages are processed
              break;
            case ChannelPowerLevel.LowPowerLevel:
              // Only Raw Push Messages are processed
              break;
            case ChannelPowerLevel.CriticalLowPowerLevel:
              // No Push Messages are processed
              break;
          }
          break;
        }
    }
  });
}
```

The `ChannelErrorType` is returned in the event arguments and can be used to determine the error types. These error types are detailed in Table 10.7.

The `PowerLevelChanged` enumeration's values are detailed in Table 10.8.

TABLE 10.7 `ChannelErrorType` **Enumeration**

Value	Description
`ChannelOpenFailed`	The channel failed to open.
`NotificationRateTooHigh`	Too many messages are being sent to the phone in a fixed period of time.
`PayloadFormatError`	The format of the headers or XML format of the message is incorrect.
`MessageBadContent`	This is used specifically for bad information in Live Tile image URLs.
`PowerLevelChanged`	The power level of the phone has changed, which will affect the push messages that are processed. The `NotificationChannel-ErrorEventArgs` class passed into the event includes an `ErrorAdditionalData` property, which can be cast into a `Channel-PowerLevel` enumeration that indicates the level of the power.

TABLE 10.8 `ChannelPowerLevel` **Enumeration**

Value	Description
`NormalPowerLevel`	All push notification messages are being processed.
`LowPowerLevel`	Only raw push notification messages are being processed.
`CriticalLowPowerLevel`	No push messages are processed.

Where Are We?

Now that you've made it through this chapter you should be ready to interact with services across the Internet. Windows Phone is a mostly connected device, which means you should be able to reach into the cloud and not only retrieve data with different services, but also push information to your application from your own services. By doing this you can create great, connected experiences for your users.

11
The Marketplace

So you've put your sweat and tears into an amazing idea that will revolutionize the way people use their phones; or you just created a cool app that outputs sounds of human flatulence. Either way you want to share your creation with the world. How do you go about it?

What Is the Marketplace?

The Marketplace (or the Windows Phone Marketplace) is where phone users can download and/or buy applications. On the phone an app called "Marketplace" lets users browse, download, and optionally buy applications that developers have written. The Marketplace is one of the main hubs on the phone, so it encourages users to look around and find applications (as shown in Figure 11.1).

The Marketplace is segregated into several different categories of applications. The hub also shows highlighted apps that the Marketplace team decides to promote on the hub. You do not have control over whether your application shows up here or not. In addition to the Marketplace on the phone, the Windows Zune client also allows users to look in the Marketplace and browse/buy applications (as shown in Figure 11.2).

This is where you want your Windows Phone application to be showcased. Although developers can install .xap files manually for up to ten applications, for mass appeal you need to be in the Marketplace.

FIGURE 11.1 The Marketplace

FIGURE 11.2 The Marketplace in Zune

How It Works

The purpose of the Marketplace is to provide a place for your application to get the exposure it needs to be successful. For any application sales, Microsoft splits the profits for direct purchases from the Marketplace. Microsoft keeps 30% and pays you the remaining 70% of any sales you make on the Marketplace.

Before you can get in the Marketplace you have to submit your application. The flowchart in Figure 11.3 shows you the process of submitting your application to the Marketplace.

You will develop your application using the Windows Phone SDK 7.1 and get it ready for submission. Submitting your app involves uploading the .xap file as well as metadata about your application. Microsoft takes this information and performs a series of tests against your .xap file, including testing it on several devices (though the devices it uses are random per application). If your application passes, it is signed by the Marketplace (which is what makes the .xap file executable on the phone) and deployed so that it can be downloaded/sold. Before you plan to submit an application to the Marketplace, you will need a membership to the App Hub.[1]

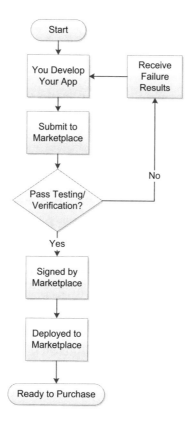

FIGURE 11.3 Submission process

---

1. You might already have an App Hub membership as it's required to unlock a device to test your application on a phone.

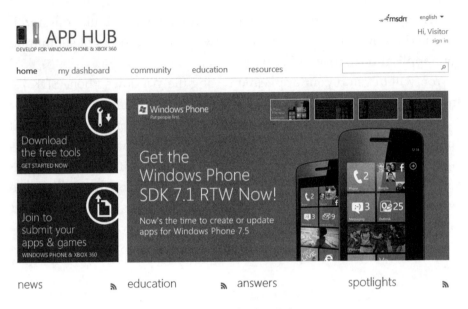

FIGURE 11.4 The App Hub

The App Hub is the gateway to your applications in the Marketplace. Figure 11.4 shows the join button on the App Hub's home page (http://create .msdn.com).

Joining the App Hub costs $99 per year for individuals or companies (students can join for free by being in the DreamSpark[2] program). By joining the App Hub you get several important benefits.

- You can submit an unlimited number of paid applications to the Marketplace.
- You can submit up to 100 free applications to the Marketplace. (Additional submissions cost $19.99 per application.)
- You can update your applications on the Marketplace for free (regardless of whether the application is a paid app or a free app).
- You can unlock up to five phones for development purposes. Unlocking a phone enables you to debug on the phone and manually install up to 10 applications.

---

2. A Microsoft student program that helps students get Microsoft development resources: https://www.dreamspark.com/.

You can join the App Hub (as an individual or as a company) on the website mentioned earlier. Membership to the App Hub is not instantaneous; before you can submit your own applications you must follow these steps.

1. You apply for membership to the App Hub.
2. You supply identity information during signup.
3. You supply a credit card number to pay the $99 fee.
4. An identity company (probably GeoTrust[3]) will contact you to confirm you or your company's identity. This information is used to ensure that people publishing apps on the Marketplace are who they say they are.

Once you have completed the signup and identity check, you'll be able to use the App Hub to unlock phones and submit applications.

Charging for Apps

If your goal is to make money on the Marketplace, you will need to understand how Marketplace sales work. You can price your app anywhere from $0.99 to $499.99 in U.S. dollars. There are fixed price points in between those two amounts, so you can't pick any random price for your application. These price points are picked purposely so that they can be converted into local currency in all the countries that the Marketplace serves. Currently the Marketplace is supported in 17 countries (though new countries may be added at any time). In these countries, the amount you charge is converted from U.S. dollars to a standard price in those countries. For example, if you charge $2.99 for an application in U.S. dollars, this would be converted into a standard price in the other markets (this is not a straight currency conversion, so the pricing will be attractively described), as shown in Table 11.1.

Fixed pricing is the only model that is available (e.g., subscription pricing is not available through the Marketplace). You cannot charge for updates to an existing application. And there are no refunds from the Marketplace.

---

3. www.geotrust.com

Table 11.1 International Pricing Example

Country	Price
Australia	AUD $4.00
Austria	EUR € 2,99
Belgium	EUR € 2,99
Canada	CAN $3.49
France	EUR € 2,99
Germany	EUR € 2,99
Hong Kong	HKD HD$25.00
India	INR Rs. 160.00
Ireland	EUR € 2,99
Italy	EUR € 2,99
Mexico	MXN $45.00
New Zealand	NZD $4.99
Singapore	SGD $4.99
Spain	EUR € 2,99
Switzerland	CHF Fr. 4.00
United Kingdom	GBP £2.49
United States	USD $2.99

When you are determining what you want to charge for an application, you also should consider the fact that users will want to try your application before they buy it. Allowing users to try your application will improve the overall sales and downloads of your application.

Your two options in this regard are to offer a "Lite" version (a free version) of your app or to support a trial mode for your app. There are benefits to both approaches. A free version will help to attract users who may only be looking for free applications (and your great app might get them to decide to purchase the app). Free apps also can get higher visibility as the top 100 apps on Windows Phone are separated into "Free" and "Paid" categories. A free version would be a separate version of the application (though probably sharing 99% of the same source code) that is limited by ads, nag screens, or limited functionality.

Supporting a trial mode for your app will enable users to download your app before they buy it. Unlike the Lite version approach, the trial mode allows users to upgrade quickly (instead of having to install the full version as a separate application). See the Using Trial Mode sidebar for more information on how to implement a trial mode version.

Using Trial Mode

Some applications in the Marketplace are available in trial versions, enabling users to try the applications before they buy them. You as the developer have programmatic access to whether a user is using the trial mode or the full version of your application through a class called License-Information. This class has a property on it called IsTrial that returns a Boolean value that states whether the app is a trial mode or full version. You can create an instance of the class and test the property like so:

```
LicenseInformation lic = new LicenseInformation();

if (lic.IsTrial)
{
    // Buy Me Nag Screen
}
```

Using this class, you can test to see if trial mode is enabled to prompt the user to purchase the application. You should cache this value in your application once it starts up, as testing this property can be resource-intensive. In fact, the Application Certification Requirements require that you cache this value or at least not call it frequently.

Getting Paid

As stated earlier, when you sell applications on the Marketplace you receive 70% of the total sales you make. Currently Microsoft is paying members of the App Hub once they break the minimum $200 threshold in a particular quarter. All payments are made via bank transfer, so you need to be able to give Microsoft banking information in your local market. The Marketplace currently allows for payments to developers living in a number of markets. In fact, Microsoft is supporting payments to developers in more markets than the number of countries that the Marketplace supports, so you can be a Windows Phone developer in quite a number of countries.

During registration you will need to give Microsoft your banking information so that the company knows how to pay you, as well as your tax information. This tax information is different for U.S.-based developers than developers outside the United States.

Tax Information for Developers in the United States

Developers residing in the United States must provide a tax identification number for Microsoft to report any earnings to. This would be a Social Security number (SSN), an Individual Taxpayer Identification Number (ITIN) for individual developers, or an Employee Identification Number (EIN) for corporations, partnerships, or associations (including nonforeign estates and domestic trusts). These tax identifiers are used to report any revenue you receive from the sale of your applications.

Tax Information for Developers Outside the United States

For non-U.S. developers the process is a little more complex. Because you are likely not considered a "U.S. Person" by the U.S. Internal Revenue Service (IRS), you must provide a U.S. tax identification number or else Microsoft is required (by law) to hold back 30% of all revenue for tax reporting. Your country may have a tax treaty with the United States so that income earned in the United States will be paid without the withholding tax. Most developers will want to apply to avoid the 30% holdback, a process that requires that you send a copy of the IRS's W-8 form to Microsoft. But this form requires an identification number. As a non-U.S. entity (person or company), you can register for an ITIN by filing a W-7 form with the IRS. Microsoft provides on its website a form letter that you can submit to the

IRS with the W-7 form. You can see a complete walkthrough of the process for non-U.S. developers on the App Hub at http://create.msdn.com/en-US/home/faq/windows_phone_7#wp7faq50.

You can also optionally submit to Microsoft a Valued Added Tax (VAT) identification number if you want to avoid being charged VAT in your country. This relates to VAT, GST, and QST (depending on what country you're in). When you supply the VAT identification number, Microsoft will send you a hard-copy tax invoice (HCTI) if that is applicable in your particular country.

When pricing your application, understand that the pricing may or may not include these taxes depending on your country. Depending on the specific country, you may be responsible for paying the taxes directly or Microsoft will remit them for you. You will want to see the current tax and payout implications for your country of origin. You can find this by going to the MSDN documentation on pricing (available via this short URL, http://shawnw.me/wp7pricing). Users can decide to pay by credit card through Microsoft billing or mobile operator billing. How quickly you are paid depends on how the user pays for your application (still dependent on reaching the $200 plateau before payments are processed).

- If the user pays by credit card (currently the majority of payments), you will get paid 15 to 30 days after billing the user.
- If the user uses mobile operator billing, you will get paid 90 to 120 days after billing.

Submitting Your App

So you have registered for the App Hub and completed the first version of your application. You are ready to get that app in the Marketplace and start sharing your creation with users. You may think you're ready to submit your creation, but before you get started some preparation is required.

Preparing Your Application

Once your application code is ready, you should perform some simple steps to make sure the submission process goes smoothly. Here is a checklist of tasks that you should complete before you submit your application.

1. Make sure your WMAppManifest.xml file is completely filled out:

```xml
<?xml version="1.0" encoding="utf-8"?>
<Deployment xmlns="..."
             AppPlatformVersion="7.1">
  <App xmlns=""
       ProductID="{12345678-1234-1234-1234-1234567890AB}"
       Title="Your App"
       RuntimeType="Silverlight"
       Version="1.0.0.0"
       Genre="apps.normal"
       Author="Your Name"
       Description="Your Description"
       Publisher="Your Company">
    <IconPath IsRelative="true"
              IsResource="false">ApplicationIcon.png</IconPath>
    <Capabilities>
      <Capability Name="ID_CAP_GAMERSERVICES"/>
      <Capability Name="ID_CAP_IDENTITY_DEVICE"/>
      <Capability Name="ID_CAP_IDENTITY_USER"/>
      <Capability Name="ID_CAP_LOCATION"/>
      <Capability Name="ID_CAP_MEDIALIB"/>
      <Capability Name="ID_CAP_MICROPHONE"/>
      <Capability Name="ID_CAP_NETWORKING"/>
      <Capability Name="ID_CAP_PHONEDIALER"/>
      <Capability Name="ID_CAP_PUSH_NOTIFICATION"/>
      <Capability Name="ID_CAP_SENSORS"/>
      <Capability Name="ID_CAP_WEBBROWSERCOMPONENT"/>
      <Capability Name="ID_CAP_ISV_CAMERA"/>
      <Capability Name="ID_CAP_CONTACTS"/>
      <Capability Name="ID_CAP_APPOINTMENTS"/>
    </Capabilities>
    <Tasks>
      <DefaultTask  Name ="_default"
                    NavigationPage="MainPage.xaml"/>
    </Tasks>
    <Tokens>
      <PrimaryToken TokenID="YourTokenId"
                    TaskName="_default">
        <TemplateType5>
          <BackgroundImageURI IsRelative="true"
                              IsResource="false">
            Background.png
          </BackgroundImageURI>
          <Count>0</Count>
          <Title>Your App</Title>
        </TemplateType5>
      </PrimaryToken>
    </Tokens>
```

```
    </App>
</Deployment>
```

Indicating the right application name, publisher, and author ensures that the WMAppManifest.xml file doesn't have any of the standard boilerplate in it. In addition, ensuring that the images for the `IconPath` and the `BackgroundImageURI` are both pointing to the right files ensures that your application isn't thrown back as having failed.

There are two `Titles` in this file. The first `Title`, in the `App` element, is used to display the title in the application list. But the `Title` in the `Tokens` element is used for the text that is displayed on the tile when pinned to the Start page. If your pinned tile image already has the title in the image, you can leave the `Title` in the `Tokens` area empty.

2. Use the Capability Detection tool to ensure that the capabilities in the WMAppManifest.xml file reflect what you think your application does.

   In the Tools section of the Windows Phone SDK, there is a tool called CapabilityDetection.exe that you can use to determine if the capabilities you're using are what you expect. The App Hub will run this tool against your main assembly to determine what capabilities are included in your application. There are some rules about certain capabilities, so you might find that your application is accidentally using capabilities you're not aware of. These capabilities are detected by what references you have (typically), so you could have an errant assembly reference to include a capability you're actually not using.

   To use the tool, open a command window and go to the Cap-Detect folder of the SDK:

```
%PROGFILES%/Microsoft SDKs/Windows Phone/v7.1/Tools/CapDetect
```

   The capabilities tool requires that you point it at the rules.xml file that is in that folder, and then point it at the main .dll of your application (not the .xap file):

```
CapabilitiesDetection.exe <<Path to Rules XML>> <<Path to App DLL>>
- e.g. -
CapabilitiesDetection.exe rules.xml c:\myapp\bin\release\myapp.dll
```

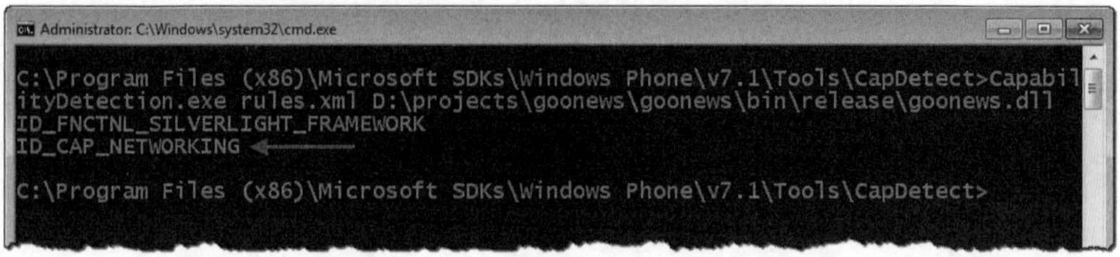

FIGURE 11.5 **Capability detection results**

You should use the included rules.xml file (as shown above) as that is what detects the specific capabilities for you. When you run the command it returns with the capabilities the tool detected. You can use these to replace the capabilities in your WMAppManifest. xml, as shown in Figure 11.5.

3. Make sure your SplashScreenImage.jpg file is in your project and it's not the default splash screen.

Each application submitted to the App Hub is required to show a splash screen image. All the templates include a default image for the splash screen. You should change this image to something more appropriate, such as a page with the application logo or starting screen. It is not required that you change it, but your application will look rushed and sloppy with the default splash screen image. This image must be a JPEG (.jpg) file and be formatted to be 480 pixels wide by 800 pixels tall.

4. Write the description of your application.

Having the description of your application prepared ahead of time means you won't rush and write a substandard description (which users can read). This description is going to be the calling card of your application, so put effort into making the description a good pitch for your product. The App Hub won't remember this information when you make updates, so keeping it as a text file in your project will mean you won't have to find the old text every time you update the application.

5. Prepare the required images.

When you submit your application to the App Hub, you'll be required to submit a number of images that are used for the

Marketplace and for people browsing applications using Zune. These include the images described in Table 11.2.

It is a common mistake to think that the background image file used for the pinned tile image (in WMAppManifest.xml) can also be used for the large Marketplace mobile icon. This is only true if the background image does not have transparency. The App Hub will accept it, but the image will look really bad in the Zune client.

TABLE 11.2 Application Images

Description	Location	Size	Format	Notes
Application icon	In the XAP as content	62 x 62	PNG	Named in the WMApp-Manifest.xml file. Transparency is allowed.
Pinned tile image	In the XAP as content	173 x 173	PNG	Named in the WMApp-Manifest.xml file. Transparency is replaced with the user's accent color.
Splash screen	In the XAP as content	480 x 800	JPEG	Must be named "Splash-ScreenImage.jpg".
Small Market-place mobile icon	Uploaded during submission	99 x 99	PNG	Don't use transparency.
Large Market-place mobile icon	Uploaded during submission	173 x 173	PNG	Don't use transparency.
Large PC icon (Zune)	Uploaded during submission	200 x 200	PNG	Don't use transparency.
Background for Zune	Uploaded during submission	1,000 x 800	PNG	Optional. Don't use transparency.
Screenshots	Uploaded during submission	480 x 800	PNG	Optional. These must be actual screenshots. You can have up to eight screenshots (all portrait).

6. Ensure that you test the application in both a light and dark theme.

    Not testing your application against both phone themes is a very common reason to get rejected by the App Hub. For example, a simple app with a background image may work perfectly fine in the dark theme (the default), as shown in Figure 11.6.

    But when installed on a phone with the light theme turned on, this app will not be usable, as shown in Figure 11.7.

    Figure 11.7 shows the light theme, but since the background is fixed, the automatic changing of the text color causes the black text to be unreadable. Test your image in both cases to be sure it works in both themes.

7. Build the project in the Release build.

    When you build the project in the Release build, you should make sure that any debug information is not included. You should also test this version on at least one real device to make sure the application

FIGURE 11.6 Works in the dark theme

FIGURE 11.7 Does not work in the light theme

works the way you want it to on a real device. The emulator is not a phone, and you don't want to get bitten by differences once you submit your application.

Once you've prepared your application, you're ready to submit it!

The Submission Process

To get started in submitting your application, go to the App Hub and log in to your account. In the App Hub you will see a "my dashboard" drop-down menu. Pick the Windows Phone option to take you to the list of your applications, as shown in Figure 11.8.

On the dashboard you'll see a list of all your submitted applications (though for your first application this list will be empty). To begin submitting your application, click on the "submit a new app" link, as shown in Figure 11.9.

This will launch the App Submission Wizard, which will walk you through the five steps of submitting your application. You can see this first step in Figure 11.10.

In step 1 you need to specify the name of the app and pick whether you want to distribute the application in the public Marketplace or as a private beta test. (See the Private Beta Testing sidebar for more information on beta testing your application.) Next, click the Browse button and select the

Figure **11.8 Accessing your "dashboard"**

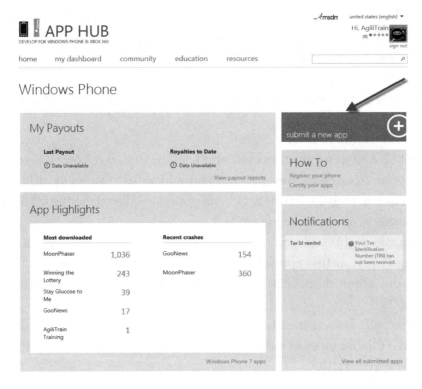

FIGURE 11.9 Starting the submission process

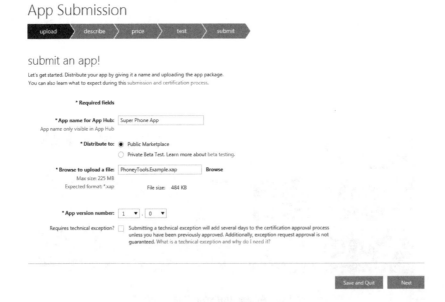

FIGURE 11.10 Step 1 of the submission process

.xap file that contains your application. This is usually in the Bin/Release/ directory of your project. Finally, pick a version number for your application. This number is used when displaying your application and should be incremented as you update your application. Once you've completed this part of the process, click the Next button to continue to the next phase (see Figure 11.11).

Private Beta Testing

When using the AppHub to submit your application you have the option to publish it to a set of users for beta testing. You can do this on the first page of the App Submission page, as shown in Figure 11.10.

By choosing private beta testing, you can invite up to 100 users (via their LiveIDs) to try your app before you open it to general availability. These users do not have to have developer devices to test your application. The rest of the process is identical to Marketplace delivery of your application. Once your beta test is over, you can move your application to the Marketplace. You will need to contact your beta testers with a link (provided by Microsoft) and test instructions to allow them to install the application on their phones.

The next step in the submission process is to include the descriptive fields for your application. This includes the category, description, and keywords. The Detailed Description section is where you should put the description you crafted in the previous section of this chapter. The Featured App description is a title that is used when your application is featured on the phone or on Zune. The keywords are used as search terms to help phone owners find your application, so make sure you fill those in. The legal URL and the support email are used to give users links to both of these items. Next, you're going to supply the images you created in the previous section of this chapter. Clicking on the box next to the area for the image will allow you to upload the image and will show you a preview of it. Clicking the Next button will take you to the pricing and distribution part of the process, as shown in Figure 11.12.

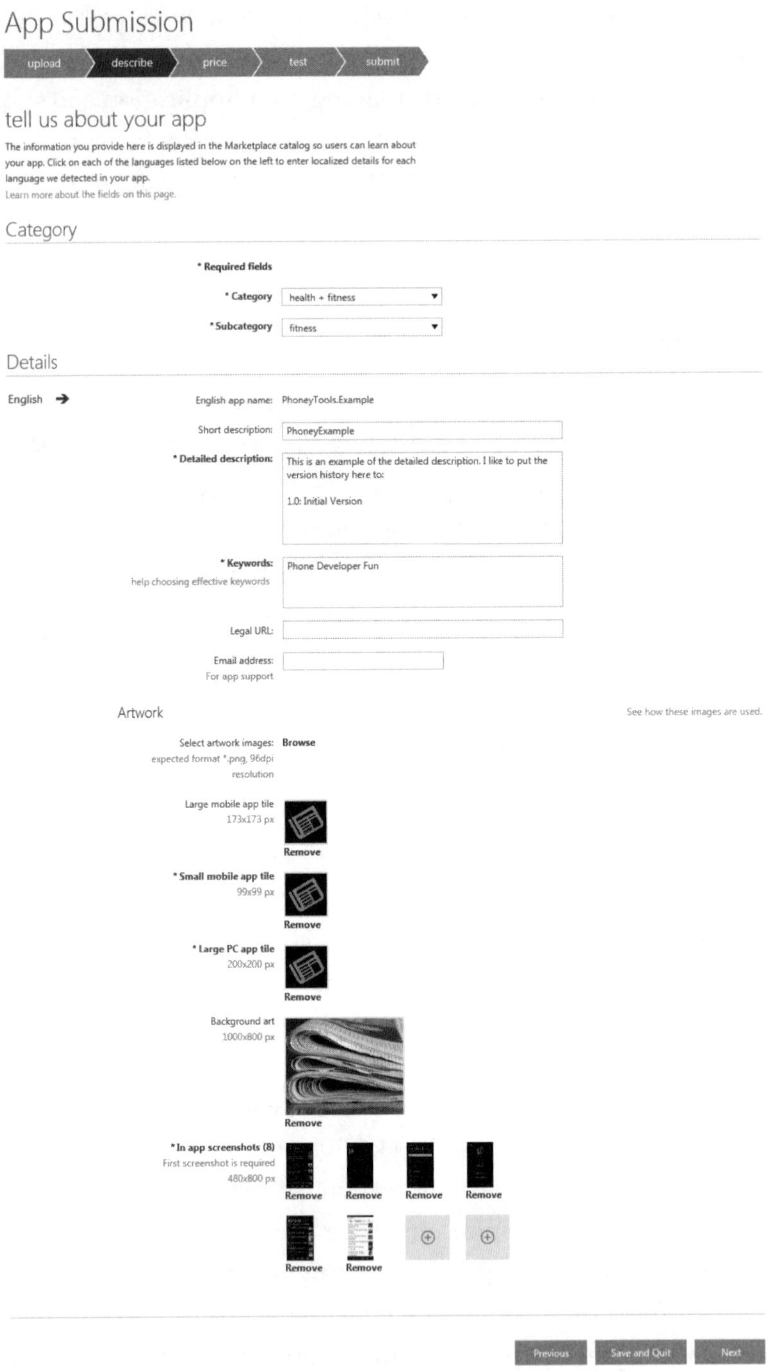

App Submission

upload describe price test submit

tell us about your app

The information you provide here is displayed in the Marketplace catalog so users can learn about your app. Click on each of the languages listed below on the left to enter localized details for each language we detected in your app.

Learn more about the fields on this page.

Category

_____

* Required fields

* Category   health + fitness ▼

* Subcategory   fitness ▼

Details

_____

English ➔

English app name: PhoneyTools.Example

Short description: PhoneyExample

* Detailed description:
This is an example of the detailed description. I like to put the version history here to:

1.0: Initial Version

* Keywords: Phone Developer Fun
help choosing effective keywords

Legal URL:

Email address:
For app support

Artwork   See how these images are used.

Select artwork images: **Browse**
expected format *.png, 96dpi
resolution

Large mobile app tile
173x173 px

Remove

* Small mobile app tile
99x99 px

Remove

* Large PC app tile
200x200 px

Remove

Background art
1000x800 px

Remove

* In app screenshots (8)
First screenshot is required
480x800 px

Remove   **Remove**   **Remove**   **Remove**

Remove   **Remove**   ⊕   ⊕

Previous   Save and Quit   Next

FIGURE 11.11 **Filling in the descriptive fields**

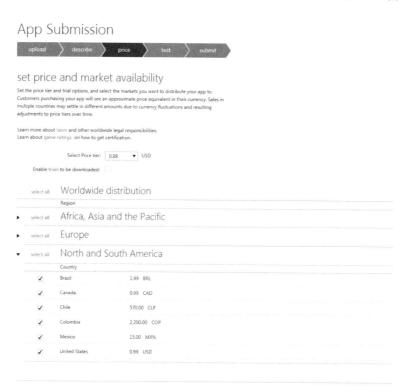

FIGURE 11.12 **Pricing your app**

In this part of the submission process you specify the pricing informa-tion. You can specify if the application supports a trial mode (you're using the `LicenseInformation` class). You can also specify whether the app is available worldwide (or at least in all of the places the Marketplace sup-ports) and what the primary offer currency is in. When you specify the cur-rency and price, you'll see the prices that will be used in the other markets. Note that you can specify (as shown) specific countries or just "select all" as is usual. Clicking the Next button again will take you to the testing and certification part of the process, as shown in Figure 11.13.

This part of the submission process allows you to add any test notes to the application (e.g., test usernames and passwords if necessary) as well as pick the publish options. You must pick a publish option. The pub-lish option allows you to specify when the application is pushed to the

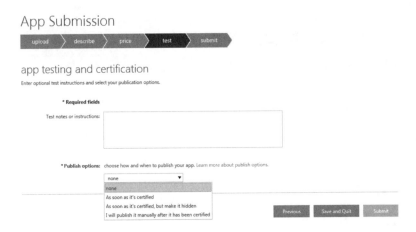

FIGURE 11.13 Publish and testing options

Marketplace. The three options are to push the application as soon as it's certified, push it as soon as it's certified but do not make it public (you will be able to make it public when you're ready), or publish it manually after certification. The publish options that do not immediately publish the application are useful because if you've submitted the application for certification but you find a bug during that process, you won't accidentally let a buggy version onto the Marketplace.

When you're ready you can click on the Submit button to submit your application for testing. If at any point in the process you want to save your place but not submit for testing, you can select Save and Quit. Once your application has been submitted, you are taken to the confirmation page, as shown in Figure 11.14.

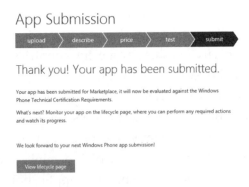

FIGURE 11.14 Submission confirmation

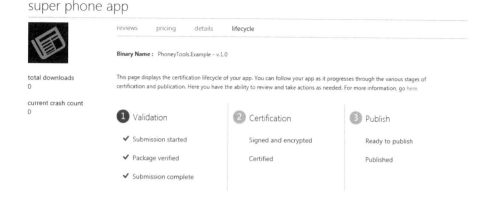

super phone app

reviews   pricing   details   **lifecycle**

Binary Name :   PhoneyTools.Example - v.1.0

total downloads
0

This page displays the certification lifecycle of your app. You can follow your app as it progresses through the various stages of certification and publication. Here you have the ability to review and take actions as needed. For more information, go here.

current crash count
0

① Validation

② Certification

③ Publish

✓ Submission started   Signed and encrypted   Ready to publish

✓ Package verified   Certified   Published

✓ Submission complete

FIGURE 11.15 Application lifecycle page

This final page shows you that your application has been submitted. You can click on the "View lifecycle page" button to see a page that shows the current status of the application (in the testing cycle), as shown in Figure 11.15.

After the Submission

After you submit your application, the dashboard will become a pretty regular page to visit as you check to see the status of your application(s). Clicking on the My Apps link on the dashboard will take you to the application status page, as shown in Figure 11.16.

The status messages on this page mean different things. Here are the statuses that your application can be in and what they mean.

- **Submission in progress:** You've started a submission but have not finished it (e.g., clicking the Submit for Certification button on the last page). This means you can continue to edit the submission.
- **Signed and encrypted:** Your application is queued to be certified by Microsoft.
- **Certified:** Certification succeeded and Microsoft is creating a signing certificate for your application to prepare it to go to the Marketplace.

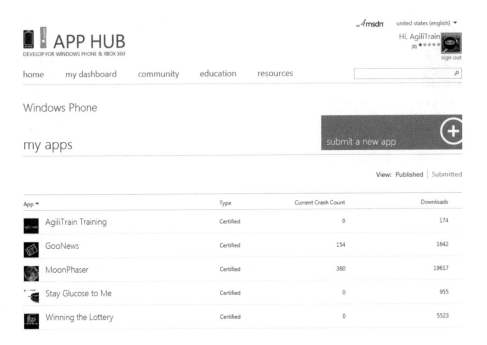

FIGURE 11.16 My Apps page

- **Ready to publish:** Your application can be pushed to the Market-place now. You can choose to publish the application whenever you are ready.

- **Published to Marketplace:** The application has been published to the Marketplace and is available for purchase. This status is indicated before the application shows up in the Marketplace. There is a pro-cess flow to get in the Marketplace that is usually less than six hours.

- **Certification failed:** The certification process failed and you can retrieve a report of why your application failed.

Once your application has been submitted, you can't make any changes or cancel the certification. The wheels are in motion, and until the applica-tion is certified or fails to certify, there is nothing you can do with your application.

While the details show a summary of the information you submitted with the application, there are two parts of this page to focus on. Near the top is a drop-down control from which you can pick operations against

FIGURE 11.17 Deep link

the application, including removing it from the Marketplace, updating the application, and submitting a different version of the app (a version of the application for another language). In the middle of the page is a deep link example. This link is a URL you can share to help people find your application. It is a Zune link to allow people to buy the application from a single link, as shown in Figure 11.17.

Using this deep link you can start marketing your app wherever you can (on your blog, Twitter, Facebook, etc.). Then the money will start rolling in!

Modifying Your Application

When you finally get your application out to the Marketplace you may find a bug, decide on a brand-new feature, or just want to change the price of your application. The App Hub makes this process simple. On the application details page, on the "lifecycle" tab, there are a list of actions near the bottom of the page, as shown in Figure 11.18.

Each action allows you to change your application on the Marketplace. The different actions are detailed as follows.

- **Hide application:** This temporarily removes the application from the Marketplace so that users cannot find it or install it. It does not remove it from devices that have already purchased the application.
- **Submit an update:** This allows you to submit a new version of the application. The update will be recertified and then moved to the Marketplace. Any existing users will be prompted to update the application through the Marketplace application on the phone.
- **Edit catalog details:** This allows you to edit the catalog details without submitting an update to the application. You can edit

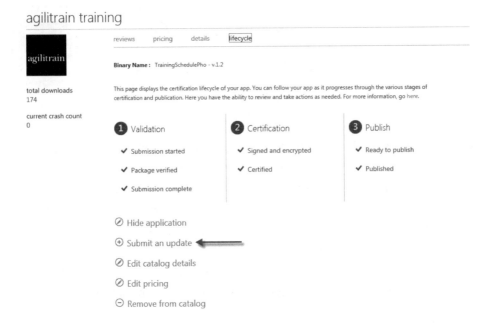

FIGURE **11.18 Application actions**

descriptions and images without requiring a complete recertification of the application.

- **Edit pricing:** This allows you to change the price of your application. This does not require a new certification. You can change your pricing at any time. You cannot move your application from a paid application to a free application. But you can take a free application and make it a paid application. Moving an application from free to paid means that every user who already downloaded the free version becomes the owner of the paid app without paying.

- **Remove from catalog:** This removes your application from the Marketplace, but current users of your application will be allowed to continue to use the application.

Dealing with Failed Submissions

If your application submission fails, Microsoft attempts to give you as much information about the failure as possible so that you can fix it. You

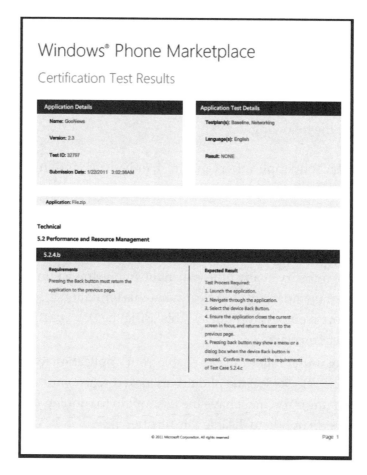

FIGURE 11.19 A failure report

can retrieve the failure report from the website, or you can have it sent to the email of record for the App Hub account. You can see a sample of a failure report in Figure 11.19.

The process of certifying your application consists of testing the way the application functions (e.g., does it run or lock up) and testing it against the Windows Phone Application Certification Requirements.[4] These certification requirements include stability, performance, resource consumption, and content requirements. An application could be rejected for a variety of reasons based on the requirements. If your application is rejected,

---

4. http://shawnw.me/wp7certreq

Microsoft gives you a detailed report of why it failed and what section of the certification requirements the failure is related to. It's important to read the Application Certification Requirements document, as there are a lot of reasons that failures occur. It is an actively changing document, but some of the major pain points that seem to bite developers submitting applications include the following.

- **Stability:** Your application can't crash or hang. (Handling `Application.UnhandledException` is important.)
- **Launch:** Your application must show its landing page within five seconds. It must be responsive to user input within 20 seconds.
- **Memory consumption:** Your application cannot take more than 90MB of memory at any time (when the phone has a minimum of 256MB of memory). It's not clear what this limitation is for phones with more memory. You can test this using the `DeviceExtended-Properties` class.
- **Running under lock:** You can enable your application to run while the phone is locked (if the phone was locked when your application was running). You must have the user opt in to running under lock and have an option to disable this functionality.
- **Back button:** You need to be careful about the navigation pattern and make sure that "Back" always takes you to the previous page and that "Back on the landing page" exits the application.
- **Using push notifications:** If you're going to use any push notifications you must tell the user when you first enable this capability and allow the user to disable it in the UI (usually via a Settings or Options page).
- **Size:** The maximum size of a .xap file is 225MB.
- **Age requirements:** If your application enables person-to-person communication (e.g., chat, IM, SMS) you must have a way to verify the user is more than 13 years old.
- **Location service:** You must allow the user to enable or disable use of location information. You also must have a privacy policy to let the user know how you're using the location information.

- **Personal information:** If you publish or share any personal information (e.g., photos, phone numbers, contacts, SMS, or browsing history) you must make the user opt in to use the functionality.
- **Content:** You cannot use nudity, sexual content, violence, or hate speech in your applications.

Using Ads in Your Apps

While you've read that you can charge for your applications and that this can help you make money, it's also fairly common to create free applications that contain advertising instead. There are several options out there that all have their plusses and minuses, as shown in Table 11.3.

The big players in in-app advertising are shown in the table, but there are some details that you'll need to determine before you decide which platform to use. By far Microsoft's pubCenter is the most attractive as it pays per impression with bonuses for click-throughs. Also, the SDK is fairly drop-in usable (though there have been reports of crashing problems). Although this is the most profitable of the advertising choices, it

TABLE 11.3 Advertising Vendors for the Phone

Vendor	URL	Model	Markets	Notes
pubCenter	http://pubcenter .microsoft.com	Per impression	U.S. only, international in future	Supports a phone SDK
AdMob	www.admob.com	Per click	International	No SDK, but simple API
Millennial	http://developer .millennialmedia .com	Per click	International	Looks like they only accept U.S. developers
Smaato	www.smaato.com/	Per click	International	Supports SDK; aggregates 50+ ad networks for high fill rate

only works for U.S. publishers. So your applications won't have advertising for non-U.S. users.

The other players all use a per-click model (which means you only get paid if the user clicks on the ad). This generally means less money, but these players do offer ads internationally and most support paying developers who are outside the United States. So there is no magical best offering for advertising inside your application.

Where Are We?

At this point you should be ready to get your application on the millions of phones out there. In this chapter you saw how to register with the App Hub, submit your application, and get it certified. Hopefully you've carefully read the checklist to make sure your application meets all the requirements to get it approved to the Marketplace. The only hard part now is coming up with that great application idea that takes the world by storm!

Index

Symbols

Safari
Books Online

FREE
Online Edition

Your purchase of *Essential Windows Phone 7.5* includes access to a free online edition for 45 days through the **Safari Books Online** subscription service. Nearly every Addison-Wesley Professional book is available online through Safari Books Online, along with thousands of books and videos from publishers such as Cisco Press, Exam Cram, IBM Press, O'Reilly Media, Prentice Hall, Que, and Sams.

Safari Books Online is a digital library providing searchable, on-demand access to thousands of technology, digital media, and professional development books and videos from leading publishers. With one monthly or yearly subscription price, you get unlimited access to learning tools and information on topics including mobile app and software development, tips and tricks on using your favorite gadgets, networking, project management, graphic design, and much more.

Activate your FREE Online Edition at
informit.com/safarifree

STEP 1:   Enter the coupon code: YOHMPEH.

STEP 2:   New Safari users, complete the brief registration form.
Safari subscribers, just log in.

If you have difficulty registering on Safari or accessing the online edition,
please e-mail customer-service@safaribooksonline.com